Acknowledgments

This guide was developed in partnership with the Community Technology Foundation of California, whose generous support, collaboration, and contribution made this book a reality, and whose leadership in the community ensures that every Californian, especially underserved communities, receives full and equal access to the benefits of technology.

CompassPoint Nonprofit Services

I would like to acknowledge the people who spend their careers dedicated to the nonprofit sector.

Cristina Chan, who birthed this book with the patience that only a mother of two could have! I couldn't have done it without you.

Jan Masaoka, whose commitment to the sector, to CompassPoint, and to me is inspiring.

Roald Alexander, aka "The Technology Wizard," for his tech knowledge, which is all over this book, and for his willingness to teach this "junior wizard."

Miriam Engelberg, whose database knowledge and sense of humor were crucial to this project and to my overall well-being.

Nelson Layag and Sabrina Smith, who were instrumental in making this project happen.

The entire staff of CompassPoint Nonprofit Services for all the support an accidental techie-turned-author could ask for.

Jonathan Stein, an amazing coteacher, cocreator, and Mac guy.

Tom Battin for the first, first draft, numerous outlines, and for just being a great guy to work with.

Kristin Long, for all her work on the web portions of this book and for her support of nonprofit accidental webbies everywhere.

Vince Hyman, Fieldstone Alliance, and the Community Technology Foundation of California for the opportunity to write the book I wanted when I started my journey as an accidental techie.

All the techies, accidental and intentional, who read drafts of this book and made excellent suggestions. They include

Brad Baso
Amanda Bergson-Shilcock
Adam Bernstein
Bryan Lloyd
Sheldon Mains
PattiJo Martain
Denis Petrov
Wendy Reed
Dirk Slater

My best bud and fellow technology author Kim Hill, who keeps telling me I can do it.

Emily Ford, and PPGC, who gave me my first chance as an accidental techie.

Sally Tatnall, the best nonprofit mentor and lifelong heart friend I could ever have.

Thanks to all!
Sue Bennett

About the Authors

SUE BENNETT was working at a nonprofit in 1998 when she introduced a radical new idea—that everyone on staff should have e-mail! The next thing she knew she was a computer network administrator—an accidental techie. After several years struggling to find resources to help her succeed in her technology role, Sue joined the staff of CompassPoint Nonprofit Services, a place where she could focus on helping nonprofit accidental techies manage while keeping their sanity. Sue is also on CompassPoint's technology faculty, developing and teaching a wide variety of computer and Internet classes across California, in addition to providing internal technical support to CompassPoint staff.

TOM BATTIN has worked with nonprofits as both an information services manager and database consultant since 1980. A specialist in technology planning and database development, he is also an experienced trainer, technical writer, and network planner. He authored the *Guide to Automating Inform ation and Referral Systems,* published by CompuMentor in 1993, and is formerly director of information technology consulting at CompassPoint.

EUGENE CHAN is director of technology with the Community Technology Foundation of California (CTFC), where he brings his experience in community economic development and nonprofit impact. At CTFC, he is responsible for developing and administering grantmaking programs that use community technology to create access, equity, and social justice for underserved communities in California. He is a cofounder and board member for the Innovation Funders Network/Technology Funders Collaborative, a national philanthropy affinity group dedicated to promoting resources and best practices in community technology.

MARY LESTER is a cofounder and now executive director of the Alliance for Technology Access (ATA), a national organization that seeks to increase technology usage by people with disabilities and functional limitations. She edited *Computer Resources for People with Disabilities* (fourth edition, 2004), a nationally acclaimed compendium of resources and information about technology. She also coauthored "Access Aware: Extending Your Reach to People with Disabilities," a planning manual for community-based organizations. Mary's community leadership includes serving on the boards of the Community Technology Centers' Network (CTC-Net), the Independent Living Network, and CompassPoint Nonprofit Services.

JONATHAN STEIN, raised by a pack of feral computers since the age of three, was quick to recognize his calling. The nomadic pack's leanings toward the nonprofit sector only solidified his decision to work with Bay Area nonprofits to improve their use of technology. Having started out as an accidental techie at Big Brothers Big Sisters in San Francisco, Jonathan now works as an independent consultant with other members of Tech Underground, providing reliable and affordable technical support to nonprofits throughout the Bay Area. He also works with CompassPoint, teaching accidental techie workshops and helping to reach out to accidental techies across the country.

MIRIAM ENGELBERG wrote and drew the Planet 501c3 cartoons that enliven this book. She is CompassPoint's chief cartoonist and a member of CompassPoint's technology faculty as well, where she teaches computer and Internet workshops, and helps "accidental database designers" as a database coach. Her book of cartoons on breast cancer will be published later this year by HarperCollins.

Written by Sue Bennett, with
Tom Battin, Cristina Chan, Eugene Chan,
Mary Lester, and Jonathan Stein

Planet 501c3 cartoons by Miriam Engelberg

With a Foreword by Tessie Guillermo,
President and CEO, Community
Technology Foundation of California

The
Accidental Techie

Supporting, Managing, and Maximizing Your Nonprofit's Technology

FIELDSTONE
ALLIANCE

SAINT PAUL,
MINNESOTA

We thank the Community Technology
Foundation of California for its funding and
participation in the creation of this book and
the David and Lucile Packard Foundation for its
support of Fieldstone Alliance.

Fieldstone Alliance is committed to strengthening the performance of the nonprofit sector. Through the synergy of its consulting, training, publishing, and research and demonstration projects, Fieldstone Alliance provides solutions to issues facing nonprofits, funders, and the communities they serve. Fieldstone Alliance was formerly Wilder Publishing and Wilder Consulting departments of the Amherst H. Wilder Foundation. If you would like more information about Fieldstone Alliance and our services, please contact Fieldstone Alliance, 60 Plato Boulevard East, Suite 150, Saint Paul, MN 55107, 651-556-4500.

We hope you find this book useful! For information about other Fieldstone Alliance publications, please see the ordering information on the last page or contact:

Fieldstone Alliance Publishing Center
60 Plato Boulevard East
Suite 150
Saint Paul, MN 55107
800-274-6024
www.FieldstoneAlliance.org

Edited by Vince Hyman
Designed by Kirsten Nielsen
Planet 501c3 cartoons by Miriam Engelberg
Cover design by Rebecca Andrews

Manufactured in the United States
Second Printing, May 2006

The Accidental Techie is one of a series of works published by the Fieldstone Alliance (formerly Wilder Publishing Center and Wilder National Consulting Services) in partnership with CompassPoint Nonprofit Services of San Francisco, California. Together, we hope to strengthen the impact of nonprofit organizations and the people who work and volunteer for them as they strive to make our communities more vital and our democracy more just.

Other titles in this series include

The Best of the Board Café: Hands-on Solutions for Nonprofit Boards by Jan Masaoka

Financial Leadership for Nonprofit Executives by Jeanne Bell Peters and Elizabeth Schaffer

Library of Congress Cataloging-in-Publication Data

Bennett, Sue, 1962-
   The accidental techie : supporting, managing, and maximizing your nonprofit's technology / by Sue Bennett with Tom Battin ... [et al.] ; planet 501c3 cartoons by Miriam Engelberg ; with a foreword by Tessie Guillermo.
      p. cm.
   ISBN-13: 978-0-940069-49-7
   ISBN-10: 0-940069-49-0
   1. Nonprofit organizations--Information technology. I. Battin, Tom, 1958- II. Title.
   HD62.6.B46 2005
   004'.068--dc22
                        2005023495

CompassPoint
NONPROFIT SERVICES

CompassPoint Nonprofit Services is one of the nation's leading management consulting and training firms for nonprofits. More than 6,000 nonprofit staff and volunteers each year attend one of CompassPoint's workshops or conferences in nonprofit strategy, finance, organization development, and technology applications. Each year more than 300 nonprofits and foundations contract with CompassPoint for consulting on the above topics, as well as on boards of directors and executive transitions, and for research studies. CompassPoint frequently works in collaborative ventures linking nonprofits, government, and private philanthropy. The organization maintains offices in San Francisco and San José.

San Francisco
706 Mission Street, 5th Floor
San Francisco, CA 94103
415-541-9000

Silicon Valley
1922 The Alameda, Suite 212
San José, CA 95126
408-248-9505

info@compasspoint.org
http://www.compasspoint.org

Community Technology Foundation of California

The Community Technology Foundation of California (CTFC) is a public foundation dedicated to increasing access to and use of information and telecommunications technology by underserved communities, seniors, and at-risk youth. Founded in 1998, CTFC operates a portfolio of innovative grantmaking, initiatives, and leadership programs through an array of partnerships with nonprofits, philanthropy, and corporations. For more information visit www.ZeroDivide.org.

Contents

Foreword

This book is soon to be the bible for accidental techies and their executive directors. Funders like ourselves will consult it when reviewing grant proposals that include computers, web sites, and technology costs. It will be within arm's reach of the thousands of accidental techies who work and volunteer in nonprofits, and right next to the mouse pads of every nonprofit person who fiddles with his or her own computer.

How do I know this? I am—in spirit and by inclination—an *accidental techie*.

At one time or another, I served as *the* accidental techie at my last two workplaces, constantly exploring how technology could make work more effective or administrative management easier. From leading one of the largest minority health advocacy organizations in the nation to starting a nonprofit from the ground up, I have directly experienced how critical the role of an accidental techie truly is. And in my current role as president and CEO of the Community Technology Foundation of California, I believe that community technology leads to greater access, equity, and social justice.

Like every foundation, we support a community of nonprofits, many of which rely on accidental techies with extraordinary levels of commitment, creativity, and passion. If our field relies so much on accidental techies to make our work easier, who can accidental techies rely on to make *their* work easier?

One answer? CompassPoint Nonprofit Services.

CompassPoint, a nonprofit itself, pioneered the accidental techie concept and first recognized its unique nature as an asset of—not a liability to—nonprofits. CompassPoint's catalog of training programs, conferences, and consulting services geared toward the technology needs of nonprofits is now more important than ever.

But what we most appreciate about CompassPoint is its strategy in nonprofit technology. While other consultants and providers argue that nonprofits need many-paged technology plans before purchasing, or that nonprofits have to hire highly skilled technologists, CompassPoint knows that community-based nonprofits will do best not by discounting and ignoring their accidental techies, but by *supporting and celebrating* them.

On behalf of my fellow accidental techies, I would like to extend my thanks to the CompassPoint team that made this book possible: Jan Masaoka, Cristina Chan, Roald Alexander, and Nelson Layag. Special thanks to authors Sue Bennett, Tom Battin, Jonathan Stein, and contributing author Eugene Chan

for writing this guide. Mary Lester, executive director of the Alliance for Technology Access, provided subject matter expertise in how to make technology more accessible. Special thanks go to Vince Hyman at Fieldstone Alliance, an editor and publisher who truly understands the value of strong nonprofits and communities.

Warmly,

Tessie Guillermo
President, CEO, and Accidental Techie
Community Technology Foundation of California

 **Welcome Techies!**

Anybody involved with nonprofit management or administration today has probably crossed paths with an accidental techie. You know, the sort of person who starts out as a program coordinator, office manager, executive director, part-time volunteer—any nonprofit role you can imagine—and who, by evolution, choice, or necessity, becomes the technology go-to person at the organization. The types of people who fill the technology support and management roles at nonprofits are as varied as their skills—which is to say, from the near-superhuman to the well-intentioned but deeply befuddled.

There is something for all these types in this book:

- The neophytes who (through wondrous good fortune or penitence for past sins, or some combination thereof) have just become the office accidental techie.
- Those who, through persistence, willpower, logic, and countless sweat-soaked hours spent untangling cables and kicking print-

> Since I knew what a motherboard was, and I had a screwdriver, I became the computer "expert" in our office.
>
> — Lary Wells, Michigan League for Human Services, Lansing, MI

ers, are de facto full-fledged career techies.

- Those rare and beloved intentional techies who chose this line of work, trained for it, and love what they're doing, despite the nonprofit jargon their colleagues toss around all the time.

Welcome to all; we are your allies. Because your leaders need to understand what you do, this guide will not only help accidental techies in concrete ways, it will also help organizational leaders and funders by giving each a better understanding of the technology management challenges facing nonprofit techies.

With a particular focus on community-based, small-to-midsized nonprofit organizations (in other words—85 percent of all nonprofits!), we've made sure this book has practical advice to help you improve your operations *tomorrow*. You can use this book now and as you go forward, taking what you need at different points as you shape the role of accidental techie in your organization.

Our Goal: To Help the Nonprofit Techie Manage Organizational Technology

The primary goal of this guide is to help you, as your organization's techie, to manage and organize technology practices in your nonprofit organization. The guide is based on our practice and experience working *as* accidental techies and working *with* nonprofit techies. Most resources about managing technology systems are written for big businesses. Advice geared to the nonprofit sector is difficult to find, especially advice that fits the scale of most nonprofits. Instead, this book addresses the technology issues you will encounter no matter what your job title is or what size organization you work for.

Rather than provide solutions that will quickly become outdated, we will provide easy-to-understand practical strategies and systems to help you manage technology and to arrive at solutions that you can use daily. This will help your organization use technology more efficiently—and, we hope, provide higher quality service, enable better planning of programs or services, and make your day-to-day life easier.

When you think about it, the impact of all of these activities is only superficially about technology. It's really about organizational capacity. The effective techie—accidental or intentional—persistently asks, How does our nonprofit pursue its mission, and how does technology assist us in doing that?

 Ten commandments for accidental techies

1. Be prepared to tackle anything involving electricity!

2. Be the guardian and protector of your organization's data.

3. Help people around you understand technology and feel friendly toward it—don't keep it mysterious or something that only you understand.

4. Learn all the ways your organization functions, and how technology currently enhances that functioning.

5. Keep abreast of how technology is used for similar functions in the nonprofit sector and in for-profit companies.

6. Support your end users. (They'll love you for it.)

7. Seek out creative opportunities to utilize technology in your organization.

8. See yourself as a catalyst for uniting users with technology to enhance the mission of your organization.

9. Be the *champion* of technology with your management and board.

10. Connect with other nonprofit techies as a means of support and information!

Help for Executive Directors

Nonprofit executives wear a thousand hats. With regard to technology, many throw up their hands in despair. Some, though, actually double as the accidental techie. Let's look at the latter first.

If you're the executive director *and* you're also the only person at your organization with any computer experience, this book will help make technology manageable by teaching you some basic **tech management skills**. This guide also will provide a template to groom the person at your organization with the second-most IT experience or interest so someone is ready to take over the tech role as your organization grows, should your own motherboard fail or you leave the organization. By having this person participate in a technology inventory, research some purchasing options, or write up some of the common step-by-step computer procedures used by staff, you may be able to ease this staff member into the job of accidental techie (instead of doing it yourself) and formalize the techie's role at the same time.

In addition, if you're the executive director and you don't have extensive computer experience, this guide will make you a better executive director. Your job is to deliver as much mission for the community's investment as possible—and technology, used well, can deliver some of the largest increases in performance a director can hope to see. So, you need to understand what that techie in your office should be doing and asking and demanding. We've come up with ten ways

> ### Accidental techie— you are not alone!
>
> Over the years, we've offered dozens of workshops for accidental techies—and how comforting it is to participants to discover that they are not alone! Accidental techies all over the country share many of the same challenges, successes, perspectives—and humor. Quotes from these techies are sprinkled throughout this book.

to help you support your accidental techie. These are big-picture goals for technology, and this book gives some practical advice on how to achieve them. What you can do tomorrow is meet with your accidental techie and talk about which of these projects is the number-one priority for your organization.

Ten things managers can do to support technology in a nonprofit

1. Recognize the accidental techie as an organizational asset. Appreciate this person!

2. Encourage formalizing the role of the technology support person.

3. Provide the accidental techie with adequate resources to evaluate your current technology.

4. Support the selection, use, and maintenance of the systems that are right for *your* organization (not for organizations in general, or for larger or smaller or different organizations).

5. Encourage employee training, for end users of technology and for your accidental techie.

6. Centralize the *management* of technology.

7. Decentralize the *support* of technology.

8. Integrate IT planning with other planning processes.

9. Get technology management on the organizational agenda.

10. Learn how to manage change and to support the accidental techie in managing change!

Help for Intentional Techies

"I'm a full-time techie in a large nonprofit. I *chose* this field! What's here for me?" If computer technology is your career—perhaps you chose technology management as your field, or you spent time in a corporate IT department before joining your nonprofit, or you've just been your organization's accidental techie long enough that it's no accident any more—there's a good chance you'll still find some info here you didn't know you needed.

For those coming to the nonprofit sector from a corporate background, we'll talk about how

---

 What we talk about when we talk about *technology*

In this guide, the term *technology* is defined broadly. It includes any technology or information technology used at a nonprofit to deliver services, track clients or members, and communicate with others—from computers, software, and assistive technology to phone systems, printers, copy machines, and postage machines.

A variety of terms and titles are used interchangeably in this guide when discussing technology and technology staff. Technology terms include

- Information technology (IT)
- Management of information systems (MIS)
- Information systems (IS)

Technology staff titles include

- Technologist
- MIS, IT, or IS staff
- Manager, director, or specialist
- Accidental or nonprofit techie

Of course, techies are also often called other names. Some of these terms are synonyms for "savior" or "genius." Others are not fit for print.

nonprofit organizations function a little differently than you may be accustomed to and how that affects technology systems. (Appendix F is particularly good for these readers.) If you're walking into an organization as its first full-time techie, lots of the systems you're used to, like a support request system or a workstation inventory, may not be there. And, oftentimes, the tools you've used in a corporate setting will be designed for a much larger scale than what's required by even a large nonprofit. You may also notice that technology users in nonprofits may differ greatly from those in the for-profit sector, especially in how they are motivated and trained. This book will describe managing IT while considering this unique culture of nonprofits.

How to Use This Guide

With so many types of readers, this guide functions at different levels. You will probably read and reread some sections, skim others, skip some sections altogether, and pull this book off the shelf for strategic answers as needed.

Whether you're new to all this or a veteran, don't let the topic overwhelm you. We've organized the guide to help you understand what technology is already in use, who's using what, how you can help them, and how you can help yourself and your organization's technology going forward. You don't have to tackle all the projects at once. Feel free to pick the areas that feel the most challenging to you or that fit your organization's needs right now.

Chapter One, What Is an Accidental Techie? explores the notion and roles of accidental techie, especially in the context of the nonprofit environment.

Chapter Two, Conducting a Technology Inventory, jumps into the first project an accidental techie probably needs to do—conduct a basic inventory of the hardware and software on hand. This important first step will make your day-to-day life less hectic immediately, and help your organization understand what it already has, and what its future needs might be.

Chapter Three, Assessing and Supporting Technology Users, helps you support your coworkers and volunteers effectively. We'll introduce you to a quick process to assess tech users and walk you through the creation of a customizable support system.

Chapter Four, Assessing and Purchasing Technology, focuses on a few of the big areas of concern for any nonprofit techie. We present a framework for purchase decision making, as well as provide some resources on managing "human capital"—consultants and volunteers.

Chapter Five, Protecting Your Organization from Disasters and Data Loss, contains the nuts and bolts of the techie's job. You'll find guidelines and recommendations on the systems your organization needs to keep information systems safe and secure.

Chapter Six, Managing Your Role in the Organization, has the tools to help you clarify your role as your nonprofit's techie—for yourself and for management. We also provide some ideas to think about for ongoing development of your technological skills.

Chapter Seven, Finding Funding for Technology, is written by Eugene Chan, director of technology at the Community Technology Foundation of California. His experience in making technology grants offers nonprofit techies, managers, and executive directors insight into what grantmakers think about when they consider funding technology. As an experienced program officer and advocate of the role of technology in improving and expanding the delivery of mission-related service, Eugene also helps other funders better understand the technology needs of nonprofit organizations.

The **Techie Tools** (appendixes), at the end of this guide, include

- The Accidental Techie's Resource Guide of web sites useful to nonprofit techies (Appendix A)

- Practical, ready-to-use templates, worksheets, and sample policies to organize your organization's technology system (Appendixes B, C, and D)

- A glossary of networking terms accidental techies should know (Appendix E)

- An article on the culture of nonprofit organizations, especially helpful for those who migrated from the business or government sectors (Appendix F)

> I am an island—a stressed and lonely island—it's just me with any real IT know-how in the agency. And mind you, I'm not a professionally trained tech-support guru, just a smart cookie who likes computers and has done database design and programming in a past life. That's along with my passion for my agency's mission. There's no budget or time to speak of and lots of problems. I'd say to potential accidental techies, "Get an agency-level IT team together with strong board or executive director backing (and some accountability to go with it)—or run for your life!"
>
> — Gale A. Shea, Family Enhancement, Madison, WI

We've also created a companion web site to this guide at www.AccidentalTechies.org, which includes

- Resources for new and experienced nonprofit techies

- Links to web sites that might be helpful in your vendor searches for software and hardware

- Links to articles of interest

- Downloadable versions of the templates and worksheets that you can customize with your organization logo and alter to fit your particular nonprofit's needs

CHAPTER ONE

What Is an Accidental Techie?

One day you heard muttering from the other room. You walked in to find the printer jammed. You unjammed the paper and saved the day. Of such simple, selfless actions, accidental techies (and other nonprofit heroes) are born. But now, somehow, all technology resources have become your responsibility!

This chapter will give you snapshots of the varied roles accidental techies play in an organization, and some of the challenges they encounter. And, of course, we can't help but give some advice along the way. (For more information on how you can manage your role as accidental techie, be sure to read Chapter Six.)

Whether you call your computers the neatest, coolest thing, Information Technology, Information Systems, or those @#*!^%$ computers, your organization uses technology and that technology doesn't run itself. In any organization—large or small—someone needs to be responsible for keeping the computers running and allowing the users to work effectively. Large organizations with

sufficient resources can hire an IT director and have specialists on staff for the network, web site, database, and user training. But many nonprofits don't have the resources to hire staff dedicated to technology. As a result, the day-to-day management of organizational technology systems becomes the responsibility of somebody who never expected to be in

charge of computers. We call that person an *accidental techie.*

A techie (accidental or intentional) is the in-house staff person who understands what the organization does and how it functions, who uses technology (most likely internal staff and possibly external clients), and, on top of all that, who can figure out how to adapt technology to the organization's goals. The accidental techie often works with external consultants who specialize in networks, web sites, software applications, or phone systems, and also makes sure that technology consultants' time is used wisely. The accidental techie does day-to-day management tasks like changing backup tapes, updating virus definitions, and clearing a printer jam, as well as documenting key procedures, tutoring individual users, and organizing staff trainings.

Accidental Techies on the Job

As an accidental techie, your job title may be receptionist, office manager, development assistant, or executive director, to name a few. If you were hired with the understanding that you'd be responsible for the computer systems, you may actually even have a job title that includes the word "technology." You may have formal training in computers, but if not, you simply may have an interest in or an aptitude for fixing things. You probably started out helping others, and as the organizational infrastructure grew, so did your responsibilities (and skills).

Regardless of their origins, most accidental techies face two unique (and confusing)

Being the office manager for a small non-profit organization means that I do a little bit of everything and ended up being the techie simply by default. It started off that I was the point person who would contact the computer support company regarding any of our problems. As time went by, our technician started teaching me how to troubleshoot problems before calling him, in an effort to save both his time and our money. The more I started to learn, the more our staff started to depend on me for problem solving.

— Brenda Bernardi, Free Arts of Arizona, Phoenix, AZ

We got new computers for the whole office three years ago. Two other staff members and I were more comfortable with computers than the rest of the staff, and we agreed to train people on Windows and MS Office. Then people started asking me questions and gradually I became the answer guy. So I had to learn more about computers so that I could fix more problems and so on . . . I like doing puzzles, and fixing computer problems is a big puzzle.

— Henry Burton, Cascadia Revolving Fund, Seattle, WA

I became an accidental techie over a natural progression of time. I am one of the youngest staff members and therefore have grown up with computers. Because I have had to use them most of my life, I think that the older staff automatically would come to me with their technology problems. As I would solve those problems, new ones would arise and because I solved the previous problems, it just became habit.

— Stacy Smith, Housing for Mesa, Mesa, AZ

questions on the job: are they a tech specialist or generalist, and what kind of actual authority do they have?

Specialist or generalist—or both?

Clearly, you don't have to become Super Techie to make your job manageable and to have your organization's computer systems run smoothly. Information technology is sometimes compared to medicine, in that there are generalists and there are specialists. As an accidental techie, you're usually a generalist. You know a bit about how the whole thing works, and you have a pretty good understanding of all the different parts and how they fit together, but you aren't able to "operate on every organ."

On the other hand, some accidental techies *are* specialists. You may have been hired because of your specific expertise in a particular fundraising software and then had your role slowly expand to include all technology-related issues. Or your hidden HTML skills (gained at a previous job) were revealed when you volunteered for the staff web site committee—thus starting you down the path to accidental techie-dom! Although you never *intended* to recommend which computers the entire organization should purchase, your specialist skills and your desire to be helpful may have resulted in your assuming more technology responsibility than you bargained for.

In any case—whether you're a specialist or a generalist and whether you love computers or wish they would all simply go away—you can be a very effective accidental techie. Just remember not to get overwhelmed by all the

PLANET 501 C3 ⸤ TALES FROM THE NONPROFIT GALAXY
BY MIRIAM ENGELBERG

WWW.PLANET501C3.ORG

NONPROFIT ACTION FIGURES

THERESA TECHIE
(FORMERLY THERESA RECEPTIONIST)

SUPER REALISTIC GLAZED OVER EYES FROM LACK OF SLEEP.

INCREDIBLE DONATED COMPUTER POWERS —
• NETWORKS MAC CLASSICS TO IBM 286'S IN A SINGLE BOUND!
• MAKES HER OWN CHIPS TO REPLACE OBSOLETE ONES!

I♡ Word-Perfect 5.1

TAP HER SHOULDER TO HEAR HER SAY: "THE FLXQRT.DLL FILE IS CORRUPT." (TRANSLATION:* "I HAVE NO IDEA WHAT'S WRONG.")

*TRANSLATOR SOLD SEPARATELY.
© 12/10/01 COMPASSPOINT

technical knowledge you don't have. The easiest way to learn what you need to know about the technology in your office is not necessarily by reading about it, but by *using* it. Sitting down and exploring the computers in your office and responding to your coworkers' cries for help can be two of the most effective methods of acquiring the knowledge that will aid you on a daily basis.

Few techies get to a place where they no longer need consultants from the outside to support systems, make technology decisions, or plan for expansion. Bottom line: *you don't have to know everything!* There are places to find help. (For instance, CompassPoint Nonprofit Services hosts two e-mail lists where you can post questions and get advice and feedback from others around the country. One is geared toward accidental techies and one toward folks in nonprofits who are working with Access databases. For more info on these services see the Nonprofit Techie Listserv and S'more Access listings under the Tips, Newsletters, and E-Mail Discussion Groups section in Techie Tools, Appendix A.)

Responsibility and authority— or neither?

Everyone in an organization interacts with technology in some way, but the person responsible for technology is rarely acknowledged as an organizational manager, and IT systems are often not institutionalized. Thus, many accidental techies face the challenge of influencing major organizational policies and procedures without real authority to do so.

Unless you're an accidental techie who also happens to be the executive director, your need to standardize systems, develop computer use policies, and purchase specialty software will most likely require authorization from others in the organization. Part of understanding your job will require you to understand your organization and your place in its decision-making structure. This question of defining and matching your level of responsibility and authority is unique to the accidental techie role; your interactions cross all staff levels and your decisions and recommendations affect the entire organization— yet you may not have any formal authority backing you up. If your role as accidental techie is very informal, Chapter Six, Managing Your Role, will guide you through dealing with this discrepancy of responsibility and authority. For now, if you've experienced this mismatch, take comfort in knowing that it's part of the accidental techie territory.

It took me a while to learn that it was okay for me not to know everything. At my last job, where I was responsible for the largest number of users and a complicated network, I often hit the wall of my own technical limitations. I learned that it was the *right* thing to do to call in a consultant. Whenever I did that, I would observe and ask questions and note things, just to informally upgrade my skills.

— Dave Moffatt, Shepherd's Center of Winston-Salem, Winston-Salem, NC

Take all the classes you can get your hands on.

— Kate Black, Sarah's Inn, Oak Park, IL

Managing Technology in a Nonprofit Organization

Some techies come to nonprofit organizations from the for-profit sector simply for the job opportunity. They don't think of IT in a nonprofit as any different from what they're used to in the for-profit sector. Then there are those techies who choose to work in the nonprofit sector because of the social impact of an organization's mission. True, a LAN is a LAN and Outlook is going to function identically regardless of one's 501(c)(3) status, but the differences in the world of IT can be significant depending on whether you are a nonprofit organization or a for-profit. Four notable areas of difference are discussed below: the atmosphere of scarcity, use of donated equipment, working with volunteers, and third-party funding.[1]

1. An atmosphere of scarcity

"We can't afford that" is a common statement at many nonprofits. Whether it's actually true or not, it is a pervasive mind-set that can affect an organization's ability to get the technology it needs. For instance, while an accidental techie may have to go through layers of management—maybe even get the board's approval—to buy a new server, you'd never think of that happening in a comparable for-profit organization.

2. Donated and nonstandard equipment

Maintaining a network of standardized computers (for example, all new computers running the same version of Windows) may be ideal, but a nonprofit may realistically depend on donated equipment and, therefore, use a mishmash of older PCs and Macintosh computers. The IT person in a nonprofit will often have to learn to work with the limited resources available and won't be able to count on consistent hardware and software

We have limited funds and spend very little on administration (less than 1 percent) so all the items we use either need to be donated, freeware, shareware, or substantially discounted.

— Tammy Van, Seven Fires Foundation, Bandon, OR

I have learned, though, that the needs of a nonprofit are different, that it's not about the newest and greatest, but what will work best in the current environment that also works with the current budget. For example, it was unrealistic to expect that all our software and hardware would be standardized, which is a common practice among large corporations. In a small organization you do not buy in bulk, sometimes you buy off the shelf. You'll have different operating systems and, most definitely, different hardware.

— Desiree Holden, Great Valley Center, Modesto, CA

upgrades. For-profit IT managers typically don't have to think about writing to a company to apply for hardware or software donations and they don't have to evaluate and accept a workstation donated by the chair of the board. More on the pitfalls of donated equipment in our inventory chapter.

At a nonprofit, the services are provided to a client and funded by a variety of sources—

---

[1] These four characteristics are derived from the article "Characteristics of Nonprofit Organizations" by Mike Allison and Jude Kaye (©1998 CompassPoint Nonprofit Services), which outlines nine areas that distinguish nonprofit organizations from for-profit entities. We've adapted aspects of this article that particularly apply to the management of technology in nonprofits. The full text of the article is included in the Techie Tools section of this guide (Appendix F).

donors, foundations, corporate donations, government funding, and the like. These contributors want to see their donations go to the program to help clients and don't always see the connection between technology and serving clients. Contributors often view technology as part of the overhead that is largely for administrative purposes rather than program uses. This means it is often difficult to raise money to purchase equipment, let alone hire staff to manage it. These limited funds mean you have to contend with outdated equipment, mixed versions of software, and donated technology.

3. Working with volunteers

If you work at a nonprofit, you may be familiar with the following volunteer scenarios: "We have this great guy who wants to volunteer for us. What can we have him do?" or "Our board member's son/daughter/brother-in-law/roommate/cousin/ neighbor is a database expert and has volunteered to write a new database for us in his/her spare time!" When's the last time you heard of a Fortune 500 company bringing in a volunteer to build a database?

Accidental techies often find themselves working with technology volunteers on technology projects. Nonprofit organizations, as a rule, rely on operational volunteers, volunteer board members, and other volunteers. (See Chapter Four for more on working with volunteers.)

In addition, you may have volunteers using your computers—to teach children, to sell theatre tickets, to connect with activist networks, or to send out notices of events. Such volunteers may not appear regularly or stay long enough to be fully trained in technology use. You'll need to design systems and procedures that allow their skills (or lack thereof!) and commitment to be maximized.

4. Third-party funding

Third-party funding—such as a government agency funding your organization to help children in foster care—is typical of nonprofits. The restrictions in a grant may limit your ability to maximize the use of that funding. In other words, you're likely to hear, "We're doing this because that's what our funder wants." For instance, many organizations struggle to find funding for general web design and development, but may have an easier time securing support to develop web-based services for a specific program area. So while the organization doesn't meet its entire goal, it receives funding that at least meets part of the goal. This may be a help (at least one part of the web site is updated) or a hindrance (the new section of the site is completely different from the rest of the site), depending on the organization's larger web site and future plans.

Following are some of the other common ways nonprofits may differ from a for-profit, large company environment, and the ways an accidental techie may experience these characteristics:

- Outdated equipment or software applications
- Systems aren't networked

- Mixed quality of equipment
- Multiple versions of software
- Limited funds that translate to no IT budget or resources for end-user training and support
- Wildly varying staff ability or willingness to use technology
- No clear management ownership of computer systems
- Donated equipment or software
- Lack of an organization-wide database
- Staff use personal e-mail accounts for work matters
- A climate of consensus that can slow down decision making
- Web sites are not updated or integrated into organization functions
- Few policies and procedures for computer systems
- Little or no documentation of systems or applications

The above discussion may lead you to think that all the characteristics of nonprofits are obstacles that make your job more difficult. That's not true! Many of the unique characteristics of nonprofits make the accidental techie job easier and more satisfying. Consider, for example:

- A value of appreciating people at all levels of the organization: Nonprofit managers often welcome reminders of the importance of technology and other support staff.

- Commitment to mission and goals: The sense of a unified team working together can create an environment that is conducive to learning and teamwork—whether about technology or about changing the world.
- Resourcefulness and ingenuity: Nonprofits are legendary for creativity and resourcefulness. Your coworkers and volunteers may be able to help you think of things you couldn't have on your own.
- Commitment to diversity and inclusiveness: Being challenged to make technology accessible to a diversity of people is more difficult *and* more rewarding than making it work for only a few.
- Connection to constituents: While for-profit companies may wish for more customer information in order to increase profits, community-based nonprofits are typically deeply embedded in their constituents and communities. As a result, your technology systems won't have to try to make people act in the best interest of your *organization*—you are trying for something much more important: helping people work together with your organization toward community goals.

"That's fine and good," you might say, "but what about the actual *computers?* How can I figure out what exactly I have on my hands and how best to deal with all these cantankerous machines?" The next chapters provide some answers to these questions to help you to make the most of this job.

CHAPTER TWO

Conducting a Technology Inventory

If you are like many nonprofit techies, you're responsible for everything that plugs into the wall. But with so many different types of technology in use, how do you know where to start? This chapter will help answer the question, What technology am I supporting? This should be the very first question an accidental techie tackles in organizing work.

This chapter will guide you through conducting an inventory of all the hardware and software that your organization currently uses, providing you with a lay of your organization's technology land. Once you have inventoried and assessed the components of your technology systems, you will be able to start building the case with leadership to support you in supporting your organization's technology.

Inventorying Your Technology

An accidental techie's first task is to figure out what equipment or hardware the organization uses and what software is on those machines. So, you will need to conduct an inventory of the following:

- Workstations
- Software
- Networks
- Internet access and e-mail
- Virus protection
- Databases
- Web sites

Workstation inventory

Creating a technology inventory can be straightforward and simple, or it can require a longer, more comprehensive overview of every technical component in your office. What's important is to capture the information that will most help you in your day-to-day work. A good workstation inventory includes

- Make
- Model
- Serial number
- Purchase or donation date
- Cost
- Warranty information
- Software installed
- Vital statistics, such as hard drive size and RAM size

You might also want to include information about the peripherals attached (printer, Zip drive, scanner, and so forth). Remember to keep a hard copy of this inventory as well as any electronic versions. It's a good practice to keep a copy off-site for the worst-case scenario!

Besides the obvious inventory items, include any assistive technology you have in your workstation inventory. The term *assistive technology* is broadly used to describe any product or piece of equipment used to increase, maintain, or improve functional capabilities of individuals with disabilities or functional limitations. This includes software as well as hardware. Some examples include screen enlargement programs, screen readers, alternative keyboards, and voice-recognition systems. See the section Accessibility, Disabilities, and Ergonomics in Chapter Three for more detailed information on assistive technology.

Software inventory

In addition to the software at each workstation, you should inventory all your software licenses, disks, and activation codes.

Keep all your licenses and disks in a secure place. Misplaced software disks are a common and easily avoidable problem.

Most current software programs require product activation codes, so it's important to track which license is associated with which workstation. While we're on the subject of software, heed this warning about pirated software. You know that program that your board chair's son-in-law loaded two years ago on your computers? Do you have a valid software license

for it? If not, don't use the program! Uninstall it from your machine. Not having proof of a valid software license could be very expensive, and there is likely a low-cost *legal* alternative (check www.TechSoup.org/Stock or check out the Open Source Software section in Techie Tools, Appendix A).

After completing the basic workstation and software inventory, you are now ready to inventory the next level of technology used at your organization. The next section covers the most common area of technology that nonprofits use—networks.

Network inventory

You need to know at least enough about your network to be able to communicate with any outside technology consultants and vendors. You need to know the type of network you have and what operating system or systems you're using. The desire to communicate with clients, funders, staff, or other organizations prompts many organizations to explore networking their office. E-mail and web sites have become essential to access and disseminate information. Many networks begin because of people's interest in sharing high-speed Internet access using DSL or cable connections. Networks are also set up to allow sharing information or resources like printers or scanners.

There are two basic types of networks: peer-to-peer and client/server. In a *peer-to-peer network* all the workstations are connected together but have equal functions and thus are "peers." In a *client/server* network at least one workstation is designated as a "server"—

Our recommendation: Strive for standard workstations

After you've completed the inventory of hardware and software at your organization, you'll probably find that everyone has something different! The equipment and programs were purchased or donated over a span of years, resulting in different operating systems, manufacturers, and software versions. That makes troubleshooting problems much more difficult.

Maintaining an office where all workstations are roughly the same make and model and use the same basic software programs has many advantages. You can know beforehand what the configuration is for a user's computer. A question about upgrades will be relevant for every workstation, and the risk of incompatibility or networking difficulties is lessened if each is roughly identical. Planning ahead to purchase as many new workstations as possible at the same time ensures that some of the hardware is exactly the same. Then you will have just a few different hardware configurations to deal with.

Donated computer equipment is a situation unique to nonprofit organizations. While it may look like a big break to a program manager whose staff are sharing one workstation, it can ambush an accidental techie. Donated machines will certainly take longer to configure for your end user than you think. Donated workstations may lack licensed software or an operating system. In fact, most people wipe all software off a hard drive before donating it. You most likely won't have any support from the manufacturer unless the computer is less than a year old. So, you will have to identify all the components of the workstation and locate all the drivers to make the donated machine function. In short, your "free" machine could end up costing you more in purchased components and staff time than the purchase price of the "brand-new, off-the-shelf-with-a-manufacturer's-warranty" machine!

Consider standardizing software. CompuMentor's TechSoup Stock offers qualifying nonprofits software at very affordable rates (see the section General and Nonprofit Specific Technology Advice, in Techie Tools, Appendix A, for more information). At minimum you should strive for

- No more than two workstation operating systems (such as Mac OS X and Windows XP, or Windows XP and Windows 2000)
- The same office suite on all computers (such as Microsoft Office XP or OpenOffice)
- The same e-mail program for everyone (such as Outlook 2003 or Eudora 6.2)
- The same Internet browser for everyone (for example, Internet Explorer 6.0 or Netscape 7.2)

The one area where this goal of standardization will not apply is assistive technology. You will need a certain degree of flexibility here, putting the technology where it is needed to serve specific employees or volunteers.

it provides "services" to the other workstations on the network.

Computers on a peer-to-peer network are connected to one another, but specific devices (like a printer) might be linked only to a single computer. However, in many cases, a network-ready printer can be connected directly to a switch or hub and be accessed from any computer on the network. The main advantage of a peer-to-peer network is that it is less expensive to set up and does not require a dedicated, separate computer to act as a server. Most operating systems (Macintosh and Windows included) have built-in peer-to-peer functions that allow you to make a printer available to other workstations throughout the network. The main disadvantage to such a network is that organizational data (such as databases or important files) is *not* centralized: each computer stores data separately, making both the sharing and backup of data difficult. So, peer-to-peer networks are best when you have a small number of computers and need to share resources like a printer or an Internet connection but don't need to share databases or other files.

> I love computers and networks when they work well, but I hate them when they don't, so I have a love/hate relationship with this part of my job. It *is* work that needs to be done and so I do it.
>
> — Karen Shain, Legal Services for Prisoners with Children, San Francisco, CA

In a client/server network, many shared resources are centralized on one server. (In some cases, with larger offices, multiple servers are needed.) The server can act as the main depot for all organization files, house (or host) the database for other computers, and generally act as the core provider of services to the entire network. While this server often runs a special version of the operating system (such as Windows 2003 Server versus simply Windows XP or Mac OS X Server versus Mac OS X), it does not necessarily have to do so. In some cases, an organization will use a computer as a server—storing central files, hosting databases—but still run a standard operating system on it. While there are several limitations to such a configuration (fewer options for security, file sharing, and print sharing, among others), it's cheaper.

If you decide on a client/server network, consider how to organize the files on the server. Basically, good file structure allows you to keep like things together, making files easier to find. Create a root folder and then whatever categories you (and others in your organization . . . this is a good time to form a team from across the organization!) decide. For example, you might want to organize your data and documents by department or service area, with different subfolders in each main category, depending on the activities of each department. Be sure to document your file structure. List or diagram all the directories (folders) and what's expected to be stored in each of them and then remember to update this diagram periodically.

Our recommendation: A local area network

In the last five years, many nonprofit organizations have embraced the joys of high-speed Internet access (via DSL, T-1 or other cable lines, or satellite) and, with it, all of the benefits of having their computers networked together. However, many small organizations still have several computers completely disconnected from each other. Thus, sharing files between staff members means e-mailing them to each other (via dial-up connection). In rare situations, some organizations have high-speed Internet access, but still do not take advantage of their internal network, or local area network.

A local area network (LAN) is a fancy term for the community created when the computers in an office are connected to each other (both the peer-to-peer and client/server networks described above are kinds of LANs). This connection can happen either through physical cables (called Ethernet cables) or through wireless technology. Regardless of how the computers are connected, a LAN allows for the simple, swift, and secure sharing of organizational files among staff members. E-mailing files back and forth or saving to floppy disk and walking down the hallway (aka the "sneaker net") to your coworker's office is not the answer. There is a better way and its name is LAN.

Technology Inventory Worksheets

One of the biggest headaches an accidental techie has to deal with is inventory. At any given time, you need to know how many computers are in use in your organization. You can elect to fill out one form per system or combine all the info into a spreadsheet or database. Sample worksheets you can use to conduct a technology inventory at your organization appear on the following pages. We've filled them in to give you an idea of how the completed forms may look. Blank copies are included in the Techie Tools and are also available for download at http://www.accidentaltechies.org.

How to use the workstation inventory record

Purchase information.

Ideally, you'll use this section to start the Workstation Inventory Record as soon as a piece of equipment arrives on your premises. Later, you'll be able to match each system on your inventory list with the invoices for those systems. You'll need the invoice information to make warranty claims.

Inventory information.

Use this section to record the model of each component of a user system. (Usually the model number, serial number, and purchase price need to be reported to your insurance company. Check with your finance director or office manager.) If you use special tags to identify company-owned equipment, record that number in addition to the component's serial number. If known, record the cost of each component. (If the cost of each component isn't known, you can simply record the total cost of the system.) In addition to recording the hardware components, you'll want to note which operating system has been installed. Don't forget to include any assistive technology that may be installed.

Network information.

If the system is connected to a network, use this section to record the type of network card, which wall jack the workstation is connected to, and the system's name on the network.

Software on this workstation.

Record all standard software and other software such as assistive technology, proprietary accounting applications or databases, as well as antivirus software and backup software if the workstation is being backed up independently from the network.

Staff who use this workstation.

Use this section to keep track of your users. If an employee leaves the company or receives a new system, record the new user's name, telephone number, e-mail address, and office location. If you decide to track users' passwords, keep them in a separate secure file.

Service and problem reports.

When the user reports a problem, use this section to record the nature of the problem. You can record additional problems on the back of the form. This section can help you identify systems or users that have particular problems.

Replacement and retirement information.

Use this section to keep track of system components that you replace. For instance, if a monitor goes bad, enter the new monitor's serial number in this section. In the Destination column, we write "trash" if we throw away the part that was replaced. We write "storage" if we keep the component (in case we can harvest replacement parts).

SAMPLE Workstation Inventory Record

Purchase information (if available)			
New: Yes	If donation, from whom:	Date purchased: 4/15/05	Accepted by: S. B.

Vendor/Invoice: *Special Computer Warehouse*

Phone: 800-745-2200	E-mail: info@SCWarehouse.com		Contact: Jeremy
Warranty: Yes	Term:		Warranty expires: 4/15/07

Tech support phone: *555-555-2200*

Comments: *Purchased through XYZ grant*

(continued)

Inventory information

Inventory Date: 4/20/05	Make/model	ID no./serial no.	Cost
__Tower			
__Laptop			
X Desktop	Compaq Evo 756	5415–85–84566	$1435.00
Monitor (e.g., Dell M90 15" CRT, 17" ViewSonic VE175 LCD)	ViewSonic VE175 LCD	5548–465–45	$456.00
Processor type/speed (e.g., Pentium4, 1G, Celeron, AMD)	3.6-GHz Pentium 4		
Hard disk size (e.g., 20 GB)	Two 160GB RAID hard drives		
Removable storage (e.g., CD-RW, CD, floppy, and/or Zip)	12X DVD+RW	52X/24X/52X CD-RW drives	
Docking station	no		
Modem __internal __external	V92 modem		
Scanner	n/a		
Local printer	n/a		
Operating system (e.g., Windows 9x, 2000, OSX)	Windows XP		
Latest OS patch (date or version)			
USB ports (number, location)	6 USB 2.0 ports, 2 FireWire ports		
Assistive technology	n/a		
Other:			

Network information

Network card (Model, driver, connection 10/100/G)	Intel PRO 1000 Gigabit NIC
Office location/network ID	Main office
Network printers attached	Training, HP4000

(continued)

Software on this workstation

Type	Name	Version	CD Key
Word processor	OpenOffice	2.0	n/a
Spreadsheet	OpenOffice	2.0	n/a
Desktop publishing	Pagemaker	7.0	555–4568–4640
Database	Exceed Basic	4.0	5654–85–9875
E-mail	Thunderbird	1.0	n/a
Web browser	Firefox	2.0	n/a
Backup	Via network		
Antivirus	Norton	2005	44654–56/52
Update method	Manual		
Assistive technology	Dragon Naturally Speaking	8.0	551526–55
Other:			

Staff who use this workstation

Date	Name	Dept.	Phone	E-mail	Office
4/21/05	Ted Franklin	Development	x 25	ted@foodforfolks.org	main

Service and problem reports

Date	Resolved	By	Details

Replacement and retirement information

Date	Part Replaced	Comments	Destination

Some other inventory tools

Belarc Advisor

http://www.belarc.com/free_download.html

This is a small program that, when installed on a workstation, builds a detailed profile of the installed software and hardware and displays the results in a web browser. The profile can be printed out and saved. The advantage to Belarc Advisor is that the information is private and kept only on your hard drive. The license associated with this product allows for free personal use only.

ENTECH Computer Inventory Software

http://epic.cuir.uwm.edu/entech/knowledge/Navigator/download.php

This program allows you to scan your computer(s) automatically and produce reports including

- Processor type and speed
- Lists of all installed programs and their versions
- Disk drive measurements including hard drive sizes, number of drives, disk space remaining, and so forth
- Network info including IP address and hostname

Best of all, this software is available free to 501(c)(3)nonprofit organizations.

TechSurveyor

http://techatlas.org/tools/

This online tool attempts to discover all the equipment in your office attached to your network. It can save time collecting information if you have a large number of networked workstations because you don't have to visit each workstation individually, at least initially. To be sure that information was accurately collected and that nothing was missed, you may still have to go around to each computer and check the data. Keep in mind that the data is sent to NPower, the organization that developed TechSurveyor. Your organization needs to decide if it feels comfortable making its data available to an outside organization affiliated with software companies.

Apple System Profiler

Both Belarc and TechSurveyor are for PCs only. If you're using Macintosh computers in your office, you can simply open the built-in Apple System Profiler (under OS 9) or System Profiler (under newer versions of OS X) to discover the main data you'll need (including hard drive size, RAM, software installed, attached peripherals, computer serial number, and so forth).

Internet access and e-mail inventory

As you inventory Internet access methods and e-mail programs, find answers to these basic questions:

- How is your organization connected to the Internet—through DSL, cable, satellite, T1 line, dial up, or some other method? Who provides the service—the local telephone company or cable company or a broker? (The local telephone company often is the provider of the actual DSL line and then you might have an Internet service provider [ISP] who supplies the connectivity to the Internet. For example, Qwest is a local phone company, and Earthlink is an ISP.)

- Do you have a contract for a particular length of time? What support is included and whom do you call if something goes wrong?

- Who hosts or "serves" your e-mail? Is there a limit to how much space you are allowed on that server?

Internet usage is often the first place people begin to consider technology as an organizational tool for program work, not just administrative work. In a strategic planning process, for example, people might start to ask, How can we use our web site or e-mail for marketing, fundraising, communication, and education?

Despite these advances, nonprofit organizations often spend little time on technology strategies. In a report titled "Appropriating the Internet for Social Change," the Social

Science Research Council says, "[Nonprofits] have not yet dipped their toe into the pool of cultural and organizational change that comes when an organization molds networked technologies into its own image, making these technologies a part of their very fabric and being." [2]

If your organization doesn't already have high-speed access to the Internet, you, as the techie, are probably feeling pressure to make that happen. A *USA Today* article estimates about 25 percent of U.S. residents live in areas where broadband is less accessible than it is in large metropolitan areas. [3] For nonprofits in such areas, slow and inconsistent dial-up Internet service prohibits the use of much more than e-mail communication. Research your options. Even if you have advanced communications services like broadband available in your area, your choices may still be few. The local telephone company or cable service may be the only provider available. You may want to investigate what public policy officials in your area are doing to improve telecommunications.

"Community wireless" is an important concept being pioneered in some communities. It means that local government makes wireless, broadband Internet access available free to everyone located in the city. Community activists may be working to bring community wireless to your town—consider joining them!

Antivirus inventory

For your inventory, you will also need to determine what kind of virus protection is in use on your computers and how it is applied to your system. Is the software installed on individual workstations? How are the definitions updated, how often, and by whom?

Our recommendation: Install antivirus software, conduct regular updates, and institute security policies

Regardless of *how* your organization connects to the Internet (and even if you're not connected), you need protection from computer viruses. E-mail viruses are the biggest threat. Viruses are computer programs that piggyback on other programs and can wreak havoc on your computer or network.

If you haven't experienced at least one devastating virus attack, you're really lucky. Hopefully you've already developed some good policies about how to prevent the release of a virus and how to protect your equipment should a virus get released.

---

[2] Mark Surman and Katherine Reilly, "Appropriating the Internet for Social Change: Towards the Strategic Use of Networked Technologies by Transnational Civil Society Organizations," Social Science Research Council, November 2003. Copies of the report can be downloaded at http://www.ssrc.org/programs/itic.

[3] Jim Hopkins, "Other Nations Zip by the USA in High-Speed Net Race," *USA Today*, January 10, 2004.

Read more on virus protection and virus protection resources in Chapter Five. A recommended checklist of factors to consider when developing a security policy for your organization is provided in the Techie Tools section.

Database inventory

You might be a database wizard, or you may never have done anything in a database but enter data. Still, as an accidental techie, you will often be responsible for maintaining databases. Keeping a hard copy of the electronic documentation of basic information about the database, like its content and purpose, will be crucial to you if the database power user (see Chapter Three for more information on power users) leaves the organization or if you have to bring in a consultant to help you with your database.

Web site inventory

Chances are your organization has a web site, or is planning to develop one soon. Some crucial information will help you manage the web site and help you interact with the consultants and vendors you rely on.

- Where is the web site hosted, and who owns the computer that houses the web site and serves pages to the public when they type in your web address? (Most organizations choose not to host [or house] their own web site, primarily because if their network goes down, their web site will remain online.) Understanding who hosts your web site, how its content can be updated

PLANET 501c3: TALES FROM THE NONPROFIT GALAXY BY MIRIAM ENGELBERG

© 10/20/02 COMPASSPOINT

WWW.PLANET501C3.ORG

I NEED YOU TO RE-DO THIS CLIENT DEMOGRAPHICS REPORT TO SHOW UNDUPLICATED CLIENTS ONLY.

CLARE DISCOVERS THE DIFFICULTY OF WORKING AT THE NATIONAL ORGANIZATION OF CLONED PEOPLE.

including the username and password, and what to do in case the site does go down is crucial for an accidental techie.

- Do you own your domain name and when is it due for renewal? What other domain names do you own?

- Who designed the web site? Who *originally* developed the web site? What tools (FrontPage, Dreamweaver) did they use? Can they show you how to make simple updates? Did they set up a username and password with the hosting company? How much attention did they pay to making the web site conform to basic accessibility standards that address the needs of people with disabilities? (See the Accessibility and Ergonomics section in Techie Tools, Appendix A, for more information on designing accessible web sites.)

Our recommendation: Well-planned and well-documented data collection systems

A database documents the flow of information through an organization. When creating a database or revising a current one, you must understand where the information originates, what form it comes in, how it is entered into the database, what happens to it there, and where it goes from the database.

Databases currently in use. Collect the following information on each database currently in use at your organization:

- What information is stored in the database? This could be a simple statement of the contents, or it could be as detailed as an entire list of tables, field names, queries, and reports.

- Who uses this database? Who has access to what parts of the database? Can everyone in the organization enter and change data, or is this restricted to database administrators? Where are the passwords stored?

- Who designed the database and what type of support is available from them? If a volunteer or consultant was hired or the database was purchased, who wrote and designed it, and can you call them? Do you have a service contract? a user manual or documentation?

- What's not working? Are there user complaints? Is data entry cumbersome? Is the database generating inaccurate reports? Does your organization need additional reports?

New databases. At its most basic, database planning is a five-step procedure:

1. Gather existing reports, and create a detailed list of reports needed.

2. Gather existing paper forms and create a detailed form list.

3. Using the form lists, create a comprehensive field list.

4. Diagram office procedures.

5. Using the field list and diagrams, create data entry screen mock-ups.

This process should result in a comprehensive list of fields, reports, and data entry screens that will serve as a wish list for the database. Label items as A (must have), B (want to have, but could work around), or C (nice to have). Prioritizing will help you find a balance between your wish list and what your budget can actually handle.

Database planning can be used to develop a new database and to evaluate databases your organization is already using. See Database Planning in the Techie Tools section for a process and sample worksheet for a detailed database plan.

For beginners: maintaining your agency domain name

A domain name is the unique name an organization reserves to link to its web site. An organization's web hosting service then "points" this domain name to the computer on which the site resides. CompassPoint Nonprofit Services, for example, must pay a fee to reserve its domain name: www.compasspoint.org. Its web hosting company then ensures that each time users type "www.compasspoint.org" into their web browser; they are "pointed" to the CompassPoint site.

Domain names must be reserved through accredited companies. (See the Web Site Resources section in Techie Tools Appendix A.) Be sure you are renewing through the company you intend to—there are huge e-mail and direct mail scams out there run by people trying to make money from organizations that don't understand how to renew their domain registration or whom to pay. The reverse has happened as well: nonprofits losing their domain names because they ignored a legitimate reminder to renew the domain name and someone else bought it. Be sure the notices are going to the currently responsible person and not to the volunteer or staff who originally registered your web site.

A written inventory of systems is crucial. Now that you've completed this first step, you'll have a better picture of what technology your organization currently has and how you use it. This makes it much easier to begin to plan future purchases, or to prioritize how to support and maintain the technology in use. These inventories can be in electronic form, but we recommend you also have hard copy printouts in a specific location in case the network goes down.

Now you're ready to learn more about your users and how to support them.

CHAPTER THREE

Assessing and Supporting Technology Users

With your inventories in hand, you know what technology your organization is using. The next step is to understand *who* is using the technology and *what* they are trying to accomplish.

This chapter will help you figure out what the technology users in your organization need. We will also introduce you to a formal help request system that will free up your time to accomplish higher-level technology planning (or to do your "real" job, if you're wearing several hats). This system will help you determine which processes in your organization are working smoothly, which users need training, and what software is malfunctioning.

The first step in this process is to inventory your users (yes, they are "your users"). They are resources as well as customers. If you invest in their development as users, you will find creative ways to help them, and they may just find creative ways to help you.

> At first most everyone was resistant to technology, but to different degrees. Once people find out how much technology can help them, some are even eager.
>
> — Denise Williamson, Western Center on Law & Poverty, Los Angeles, CA

You need to know

- What are their jobs?
- How are they organized into departments?
- Who are their supervisors?
- Which computer system are they primarily using?
- What software applications are they using?
- What are their abilities and skills?
- Do their computers need to be customized or have assistive technology added to accommodate a disability or functional limitation?

There are many ways to gather this information. You can send an e-mail questionnaire directly to users, have them complete a paper survey, or have department managers report the information. You can also collect it in conjunction with your hardware inventory. Alternatively, if staff size is small enough, you can interview users individually. While this is more time consuming, it may provide you a better opportunity to directly assess individual abilities and skills, including identifying particular needs. If you have been working with the users for a while, you may already know the answers to these questions and you may just need to compile them all into a list. You can keep this list in a spreadsheet or database. A sample is on page 33. The key is for you to learn about your users and to use the best method or combination of methods to allow you to do so. TechSurveyor also has a staff skills survey to help with collecting user data. (See the sidebar Some Other Inventory Tools in Chapter Two or the section General and Nonprofit Specific Technology Advice in Techie Tools, Appendix A, for more information on this service.)

Once collected, this data will prove to be an invaluable resource. For example, if you're planning a major software shift—upgrading existing software as well as installing completely new applications—you can quickly scan your user inventory to determine who already has experience with certain applications and who will need training and in what areas.

User Personalities

In addition to the basic who, what, why, and skill level of each user, the accidental techie may want to consider each user personality. How a user views technology can greatly influence how the individual will use your organization's technology resources. This information helps you plan how best to interact and train each user. Following are profiles of three of the most common user personalities—the savvy user, the willing user, the resistant user, and the off-site user.

Don't forget that each user is a complete human in addition to being a computer user. He may have been a computer whiz in China, for example, but he struggles with the English language. She may have seen her father lose his job to computerization, and developed a childhood resistance to labor-saving technology. Or he may be secretly a Grand Master in online strategy games, and just finds Microsoft Access too unadventurous.

The savvy user

This type of user, possibly a younger employee who grew up using computers, is generally interested in and adept at using technology. However, the savvy user also can sometimes cause problems. She can be overconfident about her skills, sometimes downloading applications she thinks are better tools than your technology team has approved, and which may in fact cause problems on the computer. But don't be discouraged.

With training, encouragement, and the occasional figurative (or literal) candy bar, the accidental techie can ask a savvy user to use her interest in technology to achieve the organization's technology goals as a departmental tech support or as a power user (see page 37 for more information on power users). This type of user might have to be reminded of the organization's technology policies and perhaps even be educated as to the purpose of the policies (see Chapter Five for more on technology policies). Many savvy users are very interested in the *why* of policies and will

SAMPLE User Inventory Spreadsheet

ID	Name	Title	DOH	Department	Supervisor
6789	Doe, Jane	Development DIR	1/1/2000	Development	Flintstone, Fred

Primary Workstation	Operating System	Non Std Apps	Std Apps	Non Std Equip
HP Pavilion	Windows 2000	Netscape	MS Office 02	

Trainings Attended	Training Needed
MS Word-Intro 5/2003	Excel and PowerPoint

ID	Name	Title	DOH	Department	Supervisor
7564	Flintstone, Fred	Executive Dir	12/15/2003	ADMIN	BOD

Primary Workstation	Operating System	Non Std Apps	Std Apps	Non Std Equip
Compaq 535	Windows 2000	Firefox	MS Office 02	

Trainings Attended	Training Needed
MS Excel-Intro 5/2004	Raisers Edge

ID	Name	Title	DOH	Department	Supervisor
5684	Wonder, Robin	Admin Assist	6/1/2004	ADMIN	Doe, Jane

Primary Workstation	Operating System	Non Std Apps	Std Apps	Non Std Equip
Compaq 535	Windows 2000		MS Office 02	

Trainings Attended	Training Needed

ID	Name	Title	DOH	Department	Supervisor
6845	Muñoz, Rosa	Program Manager	3/15/2005	Child Care	Flintstone, Fred

Primary Workstation	Operating System	Non Std Apps	Std Apps	Non Std Equip
HP Pavilion	Windows 2000		MS Office 02	

Trainings Attended	Training Needed

assist in their enforcement once they understand the need. Remember, many accidental techies started as savvy users!

The willing user

This user is willing to use technology *after* it has been proven. He won't provide any extra effort to help the accidental techie, but he will go to training if he sees how it will make him more efficient. He usually trusts that technology is a good thing, but it doesn't particularly interest him outside of work. The willing user can sometimes be converted into a departmental user support person or into a power user. This person is usually well versed in the tools he uses on a daily basis and might be willing to document procedures. This user tends to be very resistant to policies he perceives as arbitrary, but when he understands the reason for a policy, he will abide by it and is grateful to the accidental techie for his or her foresight. (And we like gratitude, don't we?)

> When working with those who are "experts," I try to call on their knowledge in areas where I'm weak.
>
> — Tricia L. Bangert, Jewish Federation of Portland, Portland, OR

The resistant user

Oftentimes, the resistant user is really just plain anxious about technology. Sometimes he or she just simply loves doing things the old way. (You may hear the resistant user mumble, "Oh, how I miss the smell of White Out in the morning. Oh, how I pine for the days of carbon paper, dial phones, and smoke-filled rooms.")

The good news is that the resistant user can be transformed! Most of us can tell stories about a user who was once terrified to touch the mouse, but who became quite proficient once given the proper instruction and some patience and encouragement. Try providing her with concrete examples of how technology will help make her job easier; for instance, teach her how to quickly search for information electronically that she currently spends lots of time to acquire manually. In some cases, folks like this may benefit from outside training—it may be easier for some people to learn from an outsider than to show one's nervousness to a coworker.

The resistant user can be especially rewarding to work with. When you help her gain confidence with technology, she can become your biggest champion. She knows firsthand the value of your time, your attention, and your ability to communicate with a variety of people.

Off-site users

It's tough to support people in off-site locations. The assistance that you can provide on-site is nearly impossible to replicate when you can't see the user's screen, particularly if the user is inexperienced or fearful. However, it can be done.

Phone support is a particular skill. Turn up your patience and reassuring tone. It helps if you know the exact configuration of the machine the person is working on and can look at a similar machine with the same configuration on your end. Software is available that allows a techie to view and sometimes control computers remotely, but be aware that some of these programs can make your computers easier to access by hackers. (See the section User Support in Techie Tools, Appendix A, for more information.)

The users of technology affect the accidental techies as much as the technology in an organization. Knowing your users' strengths, weaknesses, and customization needs is as crucial as understanding the hardware and software. This knowledge allows you to recruit a technology team and learn from each other. It's the first step in not feeling isolated in your responsibility for technology.

> I've had off-site users. They were coordinating delivery of meals to hundreds of homebound low-income elderly folks, and if they couldn't print out the map for the volunteer drivers, then people went hungry.
>
> — Dave Moffatt, Shepherd's Center of Winston-Salem, Winston-Salem, NC

> It's more difficult to work with off-site users because I can't see what they are talking about and I can't sit at their workstations and experiment. I have to depend on them describing their problems accurately and being able to follow my directions when trying to solve problems.
>
> — Denise Williamson, Western Center on Law & Poverty, Los Angeles, CA

Now that you've completed your technology and user inventories, you've got a clearer picture of what and whom you are dealing with at your organization. You're now ready to move to the next step—organizing a system to support your users to use the technology you have.

Creating a User Support System

Many people become accidental techies because of their helpful natures. You see the person in the next cubicle struggling and you offer help. You begin to get a reputation as the person around the office that knows how to lay out the newsletter or fix the printer or get an accurate report out of the database. Pretty soon, people come by your desk throughout the day expecting immediate support and you return from lunch to seventeen urgent e-mails. But how can you help everyone at once? There must be a better way. This is when instituting a user support system can help.

By creating multiple channels of support, a user support system can help minimize the number of requests that need *your* attention. Users are first asked to try to help themselves and *then* go to various others in the organization for various kinds of help. Solving simple issues on her own will develop a user's confidence, accelerate her ability to describe problems accurately, and increase awareness about how technology can be used to do other tasks. In addition, helping with technology becomes a shared function throughout the organization.

A formal support request system for the problems that *do* require the techie's attention is essential. You can prioritize those requests into the rest of your day, track the work you're doing, and keep data that can help you plan future purchases and trainings.

> We have no system in place. I think of it mostly as the "whine" system. People complain to me, and when the complaint gets loud enough, I do something about it. I know there has to be a better way.
>
> — Karen Shain, Legal Services for Prisoners with Children, San Francisco, CA

> Mostly I just respond to plaintive wails, sometimes particularly savage gnashing of teeth.
>
> — Jim Hokom, Crossroads Urban Center, Salt Lake City, UT

A multilevel support system doesn't have to be complex, but it should encourage people to first help themselves by consulting printed or online materials, then consult their colleagues, and finally, if necessary, submit the problem to the accidental techie in the manner established. Create a system that is clear, consistent, and supportive—you don't want to leave your users feeling abandoned.

Following is a suggestion for a four-level system suitable for a wide variety of nonprofits:

Level 1: User, help thyself

Level 2: Hands across the desk—Peer support

Level 3: You rang?—Formal support requests

Level 4: The light at the end—Network consultants and vendor support contracts

 I work in a five-person office. Why do I need a user support system?

Large offices aren't the only ones that can benefit from a defined IT (information technology) support structure. You may be surprised to learn that accidental techies working in even the smallest organizations have seen amazing results after establishing a very simple user support system. The system could be as simple as keeping a spreadsheet with the username, date, time, problem, and resolution. With this brief bit of information, you can prioritize requests, help track users' needs, and identify potential areas for individual or group trainings.

Level 1: User, help thyself

When confronting a technical issue, the user should first consult a "how-to" or other guide that you have developed for the organization. You probably hear a few of the same questions all the time—it will save you time and empower others to develop step-by-step instructions for these frequently asked questions (FAQs). See the sidebar FAQs: Creating "How-to" Guides for Staff, page 38, for steps to follow to begin documenting procedures and for some examples of procedures accidental techies identified as priorities for documentation.

All hardware and software manuals should be placed in a central location accessible to users. A list of web sites that users might use to research their problems might also be helpful. Although all users won't take the time to consult these extra materials, some will, and these users may even become a resource for *you*. Look through the Accidental Techie's Resource Guide in this book (Techie Tools), or visit the companion web site to this guide at http://www.AccidentalTechies.org for more resources to share with your users.

Level 2: Hands across the desk— Peer support

The next level of support in your organization should be peer support. This is probably already happening informally, as it's easy to ask the person at the next desk for help. Formalize this process.

Here's an example of how it works: James needs to reformat a Word document using bullets and automatic numbering. He is sending this document to board members before the board meeting in two days. James is having trouble with the auto-numbering function. He has already consulted, without success, the instructions from a Word training he took last summer. Instead of e-mailing the accidental techie, James first goes to Jenny, who is the person in his department designated as the Level 2 support person. Jenny is able to help James solve his problems with Word.

If you formalize a decentralized support structure (as in the above example), many requests can be handled before they even get to the accidental techie. This will allow you (the accidental techie) to concentrate on higher-level technology projects or the other aspects of your job that get interrupted every time the printer runs out of ink. Depending on the size of your organization and the level of complexity of your user support system, you may want to consider identifying "power users" as well as departmental tech support staff. Of course, this may be only one person, but oftentimes they are not.

Power users

While a departmental tech support person may be a computer generalist who can assist with many basic functions, a *power user* is someone who is highly skilled at a certain application or procedure. For instance, odds are you know who does most of the mail merges at your organization or the person who spends the most time in a certain database. Since this program staff person often will have more hands-on experience with

FAQs: Creating "how-to" guides for staff

You can save yourself a lot of time—and build your organization's tech savvy—by creating how-to guides for standard problems that staff face. Here's how to begin.

1. Once you start tracking the questions you're asked, look for patterns. Are users asking for help with the same routine tasks?

2. The next time you help someone, write down the procedure step by step. Don't skip anything. Just because you know a step, don't assume your users will!

3. Have a coworker who isn't familiar with the procedure follow your written directions. Listen to his or her feedback.

4. Change the directions according to the feedback. Remember, what seems obvious or simple to you may not be to most users (that's why they ask you for help). Save this document for future compilation.

5. As you develop more sets of instructions, compile them. (Don't try to sit down and write out instructions for everything at once.)

6. Ask your organization's management team to approve use of these instructions by everyone.

7. Distribute the manuals to all staff and volunteer users.

8. Ask that the manuals be provided to new employees as part of their orientation to your organization.

Procedures that are frequently documented in user support manuals include

- Where and how to save files
- Where and how to archive old files
- Rules about what can and cannot be installed on the desktop
- Standard passwords for everything
- How to report and log tech problems
- How to do searches on the Internet and in the databases
- How to get reports from the databases
- How to use specific types of software
- How to maneuver on the desktop and the network
- How to send an e-mail
- How to do a mail merge
- How to activate accessibility features and customize a system to accommodate disabilities and functional limitations

specialty software or certain software functions than the departmental tech support person or the accidental techie, it may make sense to make a power user the "go-to" person when anyone in the organization needs help with that particular software or procedure.

Here's an example that builds on our earlier scenario: James is ready to e-mail and snail-mail his Word document to board members. The board members' e-mail and home addresses are located in the donor database. Elena, the development assistant, uses the donor database daily. She designs her own reports and queries all the time. Elena's supervisor agrees that it's better for folks to ask "power-user" Elena for help in the database than to struggle on their own, possibly compromising the valuable information in the database, or to go to the organization's accidental techie, who is less experienced with the database. James goes to Elena for help in creating the mail merge for the board mailing. Elena shows him the most efficient way to get the results he wants, and James can ask Elena for help if something goes wrong in the future.

In another example from the same scenario: While Elena is a designated power user for the database, Jackie is the designated power user for voice recognition systems. When she suffered from carpal tunnel syndrome a few years ago, she learned how to write and edit documents using such software. Now she helps other staff who have learned to use it.

It's best to get approval from the power user's supervisor or departmental support person before formalizing the role. They'll want to know how much time their employee will spend at this task. Sometimes individuals resist taking on the added responsibility of the support role. Their supervisor can help present the idea and explain why it's useful to the organization. When approached with the right attitude, most people are receptive to helping their coworkers. Build up their confidence, explain to them why you think they are the ones to take on the task, and help them see it as professional development for themselves, and a way to help others do their best work.

> Learn to delegate. I'm still learning to do that. Everyone, especially in a small office, has to help each other. If someone has some downtime, ask them to help out. It may take a little of your time to show them what you need, but then you have a backup.
>
> — Dawn Parker, Texas Mental Health Care Consumers, Austin, TX

Level 3: You rang?—
Formal support requests

You—the accidental techie—are the next step in our multileveled support system. How do your users ask for your help now? "They just stick their head in my cube!" "I hear them shrieking across the office!"

Institute a user request system or help desk so you can end this free-for-all. This will be particularly helpful if your role as accidental techie isn't your only job at your organization, because it will allow you to schedule your response into the rest of your day. Your user request system can also help you to keep track of how you spend your time and to determine which users might need extra training or which systems are causing problems.

There are many possible request systems with just as many variations in the level of formality. Some useful methods follow:

- Keep a list on a clipboard at a corner of your desk
- Have a stack of forms that users fill out. It will only take them sixty seconds. Don't do anything without a form—it will provide documentation for developing guides and help the person making the request clarify her thoughts
- Ask users to leave messages in a dedicated voice mail box
- Set up an e-mail address dedicated to internal support requests
- Develop an online form

Keep in mind that the first system you use may not be your last. A lot depends on what works best for you and your users—if the request system is too complicated you may see a big drop in formal requests and a big increase in the number of people coming by your desk.

If all your users have and use e-mail already, it can be one of the most efficient systems to use. You will read the problem in the user's own words, and know the date and time of the request. To formalize a system, you will need to determine what level of sophistication is appropriate for your organization and

It used to be that people with problems just came to me and asked me to help fix them, whether in person, by phone, or e-mail. This worked fine in the beginning, but there are twenty-five employees in our company and we are using much more technology. It became very hard to prioritize my work, because I was constantly being interrupted with day-to-day issues. Now I use two logs. Both of these logs are in a folder on our network.

— Denise Williamson, Western Center on Law & Poverty, Los Angeles, CA

establish some type of formal procedure for your users to request your help.

Whatever kind of request form you develop, it should contain, at minimum, the following essential information:

- The date of the request
- The user's name
- The problem as reported by the user

And, after the problem is fixed . . .

- The solution—what fixed the problem (This is essential because the problem is likely to come up again and it will be incredibly helpful to be able to go back to see how you fixed it before.)
- The resources used (Did you have to research the problem on the web or call an outside vendor?)
- The time spent (How long did it take to resolve the problem?)

These request forms should be reviewed periodically to help the accidental techie and his supervisor understand and evaluate issues such as

- How is the techie's time being spent? What departments are using the techie's time? Which users have more requests?

- Do staff members need additional training in the applications or in particular functions of an application?

- Are particular systems causing problems? And, if so, do they need to be replaced?

Be consistent in how you receive and respond to requests but also make sure that requests and solutions are analyzed for the information they hold about the organization's technology.

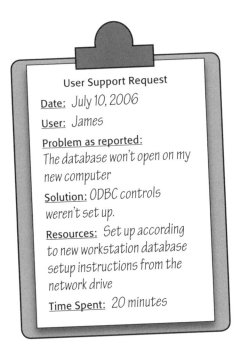

User Support Request

Date: July 10, 2006

User: James

Problem as reported:
The database won't open on my new computer

Solution: ODBC controls weren't set up.

Resources: Set up according to new workstation database setup instructions from the network drive

Time Spent: 20 minutes

Level 4: The light at the end— Network consultants and vendor support contracts

Few organizations *never* need outside technology help. An outside consultant may be hired for complex problems in areas or applications that are outside of your knowledge. For instance, assistive technology and non-English language applications are common areas in which the services of a consultant are often needed.

It's best to establish a relationship with your software vendor's tech support team and a network consultant *before* you have a problem. If you are able to plan, keep in mind that many companies offer ongoing support contracts, which can be cheaper in the long run than paying for a consultant each time you run into trouble. Contracts may be sold as blocks of time that can be used during a certain period or for a certain number of incidents or for scheduled periodic visits.

By having the accidental techie serve as the main contact for consultants, organizations can better control these costs than if every user can call in outside help. Remember that consultants are only part of the overall structure and not the first point of contact for the end user. The multilayered tech support system escalates from least expensive to most expensive, so calls to outside vendors can be limited to the problems that truly can't be solved by the resources within the organization. This saves money.

More information about working with consultants appears in Chapter Four.

Putting the System into Practice

You've provided your users with adequate self-help materials. You've set up departmental peer support or power users who take care of many of the more routine requests for help. You have management's backing for the support structure. Supervisors have agreed on who in their department will provide departmental support or act as power users. You've created easy to understand, step-by-step procedures to help users do the most routine tasks themselves, and distributed procedures and resource lists to everyone. But it will take follow-through and discipline on your part to overcome your natural urge to help and to allow the multilevel support system to function. If a user asks a routine question that you know is in the manual, you have to tell that user to consult her manual, even if you could fix it faster. If the user hasn't gone to the departmental support person or power user with an issue that you know they could help with, send the user to the proper person. This is the only way you'll free up your time to tackle some of the larger systems issues or help the users whose problems require your attention.

> What frustrates me most is that the other staff members know that I can solve the problems they encounter, or get the work done that they're supposed to do, so they just ask me without giving it much effort to learn how to do it themselves.
>
> — Jeff Campbell, Keep Austin Beautiful, Austin, TX

> As an accidental techie, I have found that you have to have several important qualities: patience, communication skills, and problem-solving capabilities. If you have those three skills and some computer knowledge (not even a lot), then being a techie really comes naturally and is oftentimes enjoyable.
>
> — Stacy Smith, Housing for Mesa, Mesa, AZ

Prioritizing Multiple Requests

It's likely that many people need help from you at the same time. The formal request allows you to prioritize which issues you address in what order. There's no right way to prioritize requests. Each situation will be individual, but we know that if a user can't get his work done, his request is a higher priority than someone who just has an irritating problem. You and your management team can set standards to ensure that issues get resolved according to the effect on the organization.

Often "workarounds" or temporary solutions might be a strategy if the solution isn't immediate. If a particular printer isn't working but other printers on the network are available to users, suggest that they use the other printer. These workarounds allow you to prioritize tasks according to the overall technology support structure rather than to users' immediate needs.

Tracking Requests and Resolutions and Evaluating the System

The formal request systems, and the filing of these requests into a log of documentation, will make your life as an accidental techie easier. It will instill users' confidence in your organization's technology and demonstrate your responsiveness to issues that arise. Keeping a written log of the issues users need help with will be a valuable tool for you, your supervisor, and your organization's planning team in anticipating future technology needs. Careful periodic evaluation of the support requests and the resources used to respond to those requests can provide information important to future planning, such as how the accidental techie's time is spent, which staff need training in what areas, and what computer systems need to be updated or replaced. This is often overlooked in the accidental techie's hurried work, but it's important. Make the time. The information you gain can end up saving you and your end users much more time than it will take to review these requests!

Through the establishment of a user inventory and the ongoing evaluation of technical request logs, you will quickly learn the skill areas of your staff as well as their gaps in experience and understanding. Once these specific areas are identified, training, whether internal or external, is often the best solution.

Staff Training

Training your coworkers or helping them find a workshop that is right for them ben- efits everyone involved. The user gains valuable skills, the organization gains a happier employee, clients and patrons experience a more efficient staff person, and you get a staff member who could be helpful to other users and who needs your help less frequently.

To figure out the various types of training your organization needs, start by looking at your user request logs. Is there a training need common to all staff? What procedures or applications have you been asked to help with most often? Make these issues the highest priority for your organization's training plan.

Types of training

One-on-one training

There are many effective ways to provide training to nonprofit technology users. Accidental techies most often provide training as one-on-one, hands-on help. The user has a question and you explain how to solve the problem. You don't always think of this as training, but you should! It's probably the bulk of the information transfer among nonprofit users.

The biggest lesson for techies when providing one-on-one training is *train, don't do.* When explaining a procedure to a user, don't grab the mouse, even though it's quicker and easier. If you do the task for someone, he or she will probably come to you again and again for the same problem. Your goal is to increase people's ability to handle most routine tasks themselves. When they become more self- sufficient, you can focus more on larger technology issues within your organization.

Troubleshooting skills

A user support system will provide a good framework for managing your users' requests. But how do you actually go about troubleshooting a coworker's particular problem? Troubleshooting involves much more than simply applying technical knowledge; it requires patience, logic, and, most important, the process of elimination. Focusing on these three very simple components is a wise strategy. Here is a process that applies this strategy.

Someone calls you to say, "The Internet is down! I've got to get this e-mail out NOW!" What do you do?

1. **Stay calm and listen.** Take a step back from the situation. Quite often, the person alerting you to a problem is agitated and has convinced himself that the sky will indeed fall if he cannot send this e-mail out within the next four minutes. Let the user vent and don't get caught up in the frenzy. Put the issue in perspective and consider the true impact of the problem on the overall operation of your organization.

2. **Ask questions.** Find out as much information as possible from the person about the problem ("I can surf the web, but I can't send e-mail") and proceed from there. Try to ask detailed questions. What the user may think is an insignificant piece of information may turn out to be an extremely helpful clue to you. The user may assume that he knows what the problem is; take time to make your own assessment before acting.

3. **Look for the obvious.** How often does the cause of a problem turn out to be simpler than you could have ever imagined? "The Internet's down" turns out to be an Ethernet cord that's been accidentally pulled out from its port. The printer isn't responding because it's turned off. Look for the easiest causes first. If those avenues produce no results, then proceed to the more intricate possibilities.

4. **Identify the problem's origination point.** Rarely will something go wrong for no reason. Therefore, one of the keys to troubleshooting is to identify the point

Small-group trainings

Small-group trainings are another effective and inexpensive way to provide technology training at your organization. When new software is installed, a new database is launched, or when you discover that many staff struggle with the same procedure, offer a lunchtime brown bag session. Small-group trainings also encourage peer support. Participants can ask each other questions based on what they all heard in the training.

The step-by-step instructions you developed for routine procedures as part of your users' self-help structure is also a form of training. Any materials you develop for small-group trainings—PowerPoint slides, handouts, and any useful sections of hardware manuals—should all be compiled as a user guide

at which everything went sour. When did the problem first arise? What changed in the situation at that point that could have caused the problem? Continuing with the scenario presented, when exactly did the user first notice that he could no longer send e-mail? Can anyone else in the office send or receive e-mail? Can everyone else access the Internet (via browser)? Each of these very straightforward, logical questions can help you get one step closer to the solution.

5. **Use the process of elimination.** This is the key to troubleshooting. Identify the problem and attempt to isolate its cause. You are told that "The Internet is down," but you learn by asking that while everyone has Internet access, no one can send or receive e-mail. This eliminates the possibility that your DSL connection is down or that there is a problem with the DSL modem. Since the entire office is experiencing the problem, you can be fairly confident that it is not simply a problem with the one user's computer or e-mail application settings. Instead, the symptoms point to a larger problem with your e-mail system. This same elimi-

nation process can be used in almost any troubleshooting scenario and, though it may seem like simple common sense, it is often overlooked as an effective means of diagnosis and resolution.

One final troubleshooting note: Don't be afraid of workarounds. If a problem arises and you're unable to determine it's exact cause, but you *are* able to identify a temporary workaround, go for it. For example, your executive director comes to you in a tizzy because she can't print to her usual color printer, which has an automatic duplex and stapling feature. She's got to get a proposal printed and sent off to a funder right away. You examine the printer and it looks like a problem that may require a few hours of work. Instead of jumping right in and attempting to fix the printer, consider your other options. If you can get it printed immediately through some other channel (another printer, taking the disk to another network, sending it to an outsourced supplier), go for it. Later, you can examine and solve the problem. In a tense, demanding situation, it's easy to overlook a temporary solution.

and distributed to all employees. These procedures should include instructions on how to customize the software using the built-in access features of operating systems. Examples include how to enlarge a cursor, adjust the size of window titles and scroll bars, magnify everything on the screen, or adjust the text and background colors to increase reading ease.

Software skills classes

Most common office applications are taught by a variety of organizations. If you're lucky enough to have a training institution in your region that is particularly focused on nonprofits, that's all the better, but even companies that focus on training corporate users can provide needed training in software for your users. Many community colleges provide courses in different software or technology skills.

Many software vendors offer classes in the advanced use of their product. The power user is often the best choice to attend a vendor-sponsored class if she is the person using a particular software program the most.

Because these outside trainings can be costly, one way to achieve maximum return is to have a user with a particular need or interest take the class. This individual can then conduct small-group or one-on-one trainings with other folks in the organization. Check the Accidental Techies web site (http://www.accidentaltechies.org) for listings of training organizations.

Other methods to provide training

Web sites can offer a wealth of information. Provide your users with lists of web sites you recommend. Check the Accidental Techies web site for ideas.

> For those afraid of or resistant to technology, I insist that they sit at their own workstation while I direct them where to click and what to do, and explain the "why" behind what they are doing. It takes a bit longer than just doing it myself, but saves me time in the long run, since they are more comfortable fixing the problem themselves next time.
>
> — Barri Waye, Fairfax Court Appointed Special Advocates, Fairfax, VA

Many online or e-learning classes are available in a variety of technology topics. E-learning works particularly well for folks who are self-directed and learn by doing. However, it often won't be effective for resistant users.

Developing a training budget

When your organization has plans for staff training, be sure technology training is part of the plan. There isn't a hard-and-fast rule about how much money should be spent on tech training. If you or your power users can perform staff trainings, you can maximize whatever training budget you have. Be sure to calculate staff time and replacement staff if you are sending operations-critical staff to training.

So where does the money come from when you need to send folks to training? Here are a couple of ideas about where training might be budgeted in your organization:

- When developing **grant budgets** for programs requiring staff to use technology, include the cost for a typical training multiplied by the number of users. For example, assume that a training costs $150 per person, and that fifteen staff people need to attend two trainings during the year. The total budget would be 15 staff x 2 trainings x $150 per class = $4,500. That $4,500 can be justified as program costs, because the program will require a particular level of proficiency from the staff that they don't presently possess.

- When planning **capital purchases**, like the purchase of new computers, add in training costs. Assume some new computer applications will be installed on new equipment and staff will need some help to get up to speed as software versions change.

Accessibility, Disabilities, and Ergonomics

Chances are some people in your organization have disabilities—disabilities that may be visible or invisible. One of the most rewarding parts of the accidental techie job is using technology to help your coworkers with disabilities, and to change the workplace so that people with all types of disabilities can work in your organization.

For us techies, it's useful to understand a person with a *disability* as someone of any age who has a functional limitation due to a disability or the process of aging and can benefit from assistive and mainstream technology. A *functional limitation* is a significant difficulty with basic life functions such as walking, breathing, talking, seeing, hearing, learning, manipulating things, or reading print.

Disabilities you may find in your organization are likely to include carpal tunnel or other repetitive stress injuries, diabetes, learning disabilities, arthritis, low vision or blindness, back problems, deafness, hearing loss, and cognitive disabilities. When you take such injuries and disabilities into account, the need for assistive technology is high!

When providing tech assistance to people with disabilities, involve them in exploring options and developing budgets. Make sure they test out equipment before purchase, and let them be the decision makers about what works best. The human resources person in your organization should be involved in setting policies around accessibility and ergonomic standards, as there are often legal considerations, and people without experience can sometimes make poor choices.

Much like troubleshooting any other technology issue, determining the right action to take when someone makes an accessibility or ergonomic request can be confusing. Here are a few basic steps to take:

1. **Listen and ask questions.** Be sure you understand what the user is expressing as his or her need. Share information about what assistive technology is available but don't assume you know what will be the best solution for the user.

2. **Document the problem** as perceived by the user.

3. **Review the list of assistive technology** provided in this chapter and research additional solutions with the user. Review the suggestions of any health care providers the user is consulting.

4. **Set up a meeting with the human resources person in your organization, the user, and yourself.** Review the possible solutions.

5. **Document the intervention** agreed upon and include it in the employee's personnel folder

Don't forget that while some people are accustomed to asking for assistance, others may be reticent, and still others may be unaware that technology can be helpful to them.

Ergonomics

When people hear the word "ergonomics," they usually start to think about keyboard placement and chair height. Ergonomics is more than that: it is defined as the application of what we know about humans to the design of objects, systems, and environments. In some organizations, a staff person's responsibilities include making sure that the workplace is safe, accessible for people with disabilities, and designed to minimize risk of injury.

But realistically, the job of computer-related ergonomics is often up to the accidental techie. And after all, because a techie's job is to help people make use of technology in support of the organization's mission, we have to think about ergonomics as being as important as well-configured hardware.

One thing that makes ergonomics difficult is that there are no one-size-fits-all solutions. Each user is unique. Sometimes a user can make suggestions as to what he needs to be more comfortable and safe, based on recom-

---

 Accessibility policies

Your organization may want to consider having a stated policy about accessibility issues. This is a sample of a policy you could adapt to fit your organization. Keep in mind that all personnel policies should be coordinated through your human resource manager and should be reviewed for legal implications.

Sample accessibility policy

The *organization* maintains a fully accessible office and conducts all business in a fully accessible manner, including all programs and services.

The *organization* will make all reasonable accommodations for new employees or those returning to work after an illness or injury. People are responsible for requesting needed accommodations.

The *organization* encourages people with disabilities and also people with family members with disabilities to apply for positions of employment at the *organization*.

The *organization* operates with a philosophy of mutual respect and strives to create an environment that is supportive for everyone. Some employees may have an environmental illness (or multiple-chemical sensitivity) and be highly sensitive to perfumes, colognes, or other scented personal care products. We ask that you consider the sensitivity of others before you choose to wear a scented product. Any employee who is experiencing a problem with a scented product worn by a fellow employee is encouraged to respectfully request that the work area be kept fragrance free.

mendations from his health professional or prior experience with hardware or software.

Rather than take on all aspects of workplace safety and ergonomics, we focus here on what is likely to be the province of accidental techies—the workstation and the computer.

The principle of universal design is that everyone—not just people with disabilities—benefits from technology that can be modified to individuals. (A classic instance of universal design is "curb cuts"—first made to allow people with wheelchairs to cross streets, it's now clear that people pushing strollers or hand trucks and many others benefit from curb cuts.)

Sticking with this principle, accessibility and ergonomics are not separate topics from the topic of disabilities. Everyone customizes their technology to some extent to meet their own needs and styles, for instance, by arranging hardware and connections for maximum convenience, organizing the desktop, adjusting the display brightness and contrast and the sound volume, and arranging files in the way that is most efficient for them. In the case of people with disabilities, the customization can mean the difference between being able to use a computer or not.

For some people it's not a question of adding different types of technology, it's about the height and position of things. There should be options for different work surface heights and underdesk clearances. The most versatile workstations have adjustable components. These include adjustable height and positioning of the tabletop, keyboards, mouses or trackballs; adjustable placement of monitors

and printers; and, if needed, adjustable wrist and arm supports. Also, lighting should be designed to maximize visual access. Brighter lighting is not necessarily better—excessively bright or poorly angled lighting can create glare that makes viewing a monitor difficult. Lighting should be adjustable.

> I know that our web site isn't accessible to people with disabilities, but I don't have the skills to address it. It's frustrating.
>
> — Sara Morrison, Massachusetts Council of Human Service Providers, Boston, MA

Accessibility technology

Most people interact with the computer using a keyboard, mouse, screen, and speakers, and rely heavily on reading and writing in that interaction. A disability may impact an individual's level of success with one or more of these elements. While we do not go into extensive detail, this section provides information on the basic types of accessibility accommodations that are more commonly used by people with disabilities and functional limitations.

There are two main categories of accessibility modifications or solutions:

1. Built-in universal design features

2. Specialized technology devices and software designed specifically for people with disabilities, commonly known as assistive technology

49

Built-in universal design features

Standard technology is now developed using the principles of universal design. Standard operating systems include built-in features and options to meet the needs of many people with disabilities. The biggest advantages to universally designed technologies are the absence of additional cost and their availability to everyone. All that is required is the knowledge of how to use them.

Macintosh and Windows operating systems each have specific features that address the keyboard, mouse, screen, sounds, and the reading and writing process. For more information on accessibility features in the Mac operating system visit http://www.apple.com/accessibility. On Macintosh OS systems prior to OS X, access these features through the Control Panel. In OS X, access them through System Preferences > Universal Access.

For more information on accessibility features in the Windows platform visit http://www.microsoft.com/enable. Windows has an Accessibility Wizard tool that guides you through customizing many of the options. Access this feature from the Start Menu, then select Programs > Accessories > Accessibility > Accessibility Wizard.

The Alliance for Technology Access (http://www.ATAccess.org) also lists resources associated with both operating systems on its web site. Two ATA resources that may be particularly helpful are the guides "Access Aware: Extending Your Reach to People with Dis-

abilities" and *Computer Resources for People with Disabilities, 4th Edition.*[4]

Following is a list of some of the accessibility features currently available in operating systems. Exciting new tools are under development that will improve the accessibility of computer systems. Use this list when discussing accessibility with one of your end users—he or she may not realize how much is available!

- *Sticky Keys* (available on Windows and Mac): Lets you hold down keys in sequence, rather than having to hold down multiple keys at the same time. Helpful for someone who is using one hand or has limited dexterity.

- *Filter Keys* (available on Windows): Instructs the computer to disregard keystrokes that are not held down for a minimum period of time, adjusts or turns off the key repeat feature, and can ignore double strikes.

- *Slow Keys* (available on Mac): Changes the length of time needed for a keystroke to register. Prevents keys from being entered accidentally.

- *MouseKeys* (available on Windows and Mac): Controls the mouse pointer with the numeric keypad instead of the mouse. Helpful for people with low vision or who have difficulty with their hands.

- *Onscreen keyboard* (available on Windows): Displays a virtual keyboard on the computer screen that enables typing using

---

[4] Alliance for Technology Access. (Albany, CA: Hunter House Publishers, 2004.)

a mouse, alternative pointing devices, or a single switch. Used by people who cannot use a standard keyboard.

- *Customizable mouse pointer* (available on Windows): Adjusts the mouse pointer to normal, large, or extra large and varies color and animation. Helpful for people with low vision.

- *Mouse Tracks* (available on Mac): Adjusts mouse tracks and thickness of the I-beam cursor.

- *Scalable user interface elements* (available on Windows): Adjusts the sizes of window titles, scroll bars, borders, menu text, icons, and other elements. Useful for people with low vision.

- *Magnifier* (available on Windows) and *CloseView* (available on Mac): Magnifies what is on the screen. Can invert the screen color (white text on black) for higher contrast. Useful for people with low vision.

- *Narrator* (available on Windows): Reads what is displayed on the screen. Works only with certain programs and utilities—very limited, but can be helpful during the installation process for people with vision, learning, or cognitive disabilities, or people using English as a second language.

- *Text-to-speech* (available on Mac): Reads the alert messages, selected text, and the text under the mouse pointer. Helpful for people with vision, learning, or cognitive

disabilities, or people using English as a second language.

- *SoundSentry* (available on Windows): Sends a visual cue such as a blinking title bar whenever the system makes a sound. Useful for people who are deaf or hard of hearing.

- *System Alert* (available on Mac): Menu bar blinks for system alerts. Useful for people who are deaf or hard of hearing.

- *Customizable sound schemes:* Adjusts sounds and volume associated with various onscreen events. Useful for people who are hard of hearing.

As you can see, both systems offer many ways to customize the action of the keyboard, the color and size of all items on the monitor, the control of the mouse, and the speech and sounds of the system.

Specialized assistive technology

Specialized assistive technology tools have typically been developed by relatively small third-party developers and vendors to meet specific needs of individuals with disabilities. Examples include keyboards with larger letters or larger keys, screen magnifiers, and screen readers. This technology helps many people who need more features and choices than are included in standard products. Below, we touch on some of the fairly common products available to accommodate someone's disability on the job. These tools range in sophistication, ease of use, and price, depending on the number of features needed.

- *Alternate keyboards.* Available in a wide variety of types, shapes, sizes, layouts, and complexity. Some are programmable. Others provide a larger-than-standard target area to accommodate a smaller range of motion.

- *Ergonomic keyboards.* Designed to provide for natural and ergonomically correct hand positions. Some also have options for varying position during the work period to reduce the likelihood of repetitive stress injury.

- *Large-print labels.* Can be stuck to any keyboard to enhance the visibility of the keys.

- *Wrist and arm supports.* Can help alleviate or prevent repetitive stress injuries. Arm supports are devices that stabilize and support arms and wrists when typing or using a mouse or trackball. Wrist rests support the wrist while using a keyboard.

- *Voice recognition.* Also called speech recognition, because users speak to the computer instead of using a keyboard or mouse. Typically used by people who have significant difficulty using their hands. Can be used to create text documents, to browse the Internet, and to navigate among applications and menus by voice. Many allow you to just navigate the computer, that is, control the cursor and execute commands by voice. The accuracy of these systems is improving, but there are still many challenges to using them in busy offices. Not recommended for people with vision impairments or learning disabilities because you need to be able to identify and correct errors.

- *Word prediction programs.* Offers word choices from a list that is generated from the first one or two letters typed. Can help cut down on key strokes and save energy. Also helpful for people with learning disabilities.

- *Abbreviation expansion programs and macro programs.* Helpful to people who have trouble with a standard keyboard or those with memory problems. Abbreviation expansion programs assign a series of letters, words, or sentences to one or more keystrokes. Macros "record" a long series of commands and assign them to a function key, combination of keys, menu item, or onscreen button. For example, with one or two keys, you could open a word processor, enlarge the text, and enter a name or address.

- *Pointing alternatives.* Devices other than the usual mouse, such as trackpads, trackballs, game controllers, and small joystick-like pointers built into the keyboard. Some electronic pointing devices control the cursor on the screen using ultrasound or an infrared beam. The mainstream market offers a wide variety of trackballs and joysticks.

- *Screen enlargement programs.* Full-featured programs that provide sophisticated control and enlargement of everything on the screen. Some include speech output in addition to magnification. Helpful for people with low vision.

- *Screen readers.* Sophisticated programs that read aloud everything on the screen. Designed for and used by people who are blind.

- *Talking and large-print word processors.* Programs that use a speech synthesizer to read aloud what has been typed. Can help people with learning disabilities or low vision.

- *Monitor additions.* Devices that enhance or alter the use of a standard computer monitor. Exterior screen magnifiers fit over the screen of a computer monitor and magnify the images that appear on the screen, but result in distortion for many. Antiglare filters are clear screens that fit over a computer monitor screen and reduce glare and improve contrast, and also reduce ultraviolet rays and other energy emissions. Monitor mounts come in a variety of styles and degrees of flexibility and allow adjustment of the monitor position.

- *Reading tools.* Software designed to make materials more accessible for people who have difficulty with reading. A common system would include a scanner and OCR (optical character recognition) software. After creating an electronic document, you can reformat the text or have it read aloud.

- *Reading and writing aids.* Programs that allow text to be read aloud as you type. These include word prediction, comprehensive spell checking features, and talking dictionaries and thesauruses. Use of auditory feedback is important in all these programs. Helpful for people with learning disabilities.

Some of the most widely used tools are common ones used by the general public. A prime example is a 21-inch monitor, which can greatly improve the ability of someone with low vision to use a computer. As another example, headphones that plug into external speaker jacks are useful if you want your computer to talk to you and not bother people around you. You can also plug in speakers or amplification systems if you need more volume. A tool does not have to be "specialized" or expensive in order to be useful to someone with a disability.

As this chapter shows, understanding staff members' technology strengths and needs and addressing them with appropriate systems and individualized responses can help them harness technology effectively. Naturally, this helps you be an effective accidental techie, which helps your nonprofit achieve its goals and mission. And that's *always* an exciting process to be a part of!

In the next chapter, we'll introduce more concepts to make your job easier—understanding how to hire outside help and how to make the decision to purchase new technology.

Assessment of organizational readiness with assistive technology

Even in the unlikely event that no one in your organization currently uses assistive technology, chances are that someone could benefit from it now, and it's all but certain that others will in the future. Reread the section on assistive technology on the previous pages and then use the questions below to survey your organization's readiness to use assistive technology.

	Yes	No
1. Do all of your computers have their operating system's accessibility features installed?	❏	❏
2. Have your staff and volunteers been trained in the use of operating system accessibility features?	❏	❏
3. Is there a staff person who as part of his or her job description is responsible for the accessibility of your computers?	❏	❏
4. Does your organization maintain a list of contact information for assistive technology resources, such as Alliance for Technology Access (ATA) members or other assistive technology providers?	❏	❏
5. Is a written plan in place for the acquisition, upgrading, and maintenance of assistive technology hardware and software?	❏	❏
6. Is a written plan in place for training new and existing staff on new assistive technology?	❏	❏
7. Does your organization have a way for computer users to notify staff of their accessibility needs?	❏	❏
8. Does your technology inventory include products from the following categories?	❏	❏
8a. Keyboard alternatives. Some people may have trouble using a standard keyboard because of dexterity or motion limitations. Do you offer alternatives to a standard keyboard such as a compact keyboard, large-key keyboard, onscreen keyboard, or voice recognition system? If yes, which? _____	❏	❏

	Yes	No
8b. Mouse alternatives. Some people with functional limitations in the use of their hands cannot use a standard mouse. Do you offer alternatives to a standard mouse such as a trackball, joystick, or touch screen?	❏	❏
8c. Enhanced view monitor. Someone with low vision who needs to magnify what they see on the screen requires a larger monitor than the standard 14- or 15-inch screen. Do you offer a monitor that is 21 inches or larger?	❏	❏
8d. Voice output software. Someone who is blind or has difficulty reading English benefits from software programs that use synthesized speech to read aloud what appears on the screen. Do you offer software program(s) with voice output capability?	❏	❏
8e. Miscellaneous low-tech aids. Many simple and inexpensive products can make a computer more accessible for a wide variety of people. Do you have large-print high-contrast keyboard labels, wrist supports, key guards, or adjustable monitor mounts?	❏	❏

CHAPTER FOUR

Assessing and Purchasing Technology

You know what technology your organization uses, who uses it, and what type of help they need. You've got the users following a procedure to request your help. Now you just need to keep it all going!

In this chapter, you'll learn about more systems and processes to help your day-to-day operations run more smoothly and to help your organization plan for future technology needs. In particular, we'll focus on these projects:

- Assessing new technology
- Purchasing technology and choosing vendors
- Managing consultants and volunteers

Assessing New Technology

"Hey, I just heard about this great new device/software/web tool that will solve all of our problems." Sound familiar? Technology

> When it comes to hardware I decide what we get. I am consulted on software, and I usually purchase and install all hardware and software. I have significant budget responsibilities. Since I enjoy technology, most decisions are either left to me, or I am part of the decision team.
>
> — Lary Wells, Michigan League for Human Services, Lansing, MI

is constantly changing, so you will be called on to help with many purchases. One mistake nonprofits make is expecting that the same company selling you equipment or applications will determine what you need—in other words, relying on the vendor to be your purchasing consultant. A vendor can provide one opinion about what you should buy, but this opinion should never be the only one. We suggest the following steps to help you make informed decisions on your organization's technology purchases. And remember, technology is just a tool to help your organization fulfill its mission.

1. Identify and clarify the want or need
2. Try to build on what you have now
3. Talk to colleagues about how they solved a similar problem
4. Define your decision criteria

5. Define your decision-making process

6. Be sure to comparison shop

1. Identify and clarify the want or need

While technology is often marketed as a way to increase efficiency, you need to know more tangibly how your organization will use a new technology. It may help folks accomplish something more easily or faster, or it may allow them to do something they can't currently do.

When first considering new technology you should ask, What is the perceived need and will the proposed technology solve it? Then ask, How else can I solve this problem? Is new technology really the solution?

Don't confuse highly functional technology with highly trained staff. Don't confuse a newer, faster workstation with the lack of an efficient flow of information. Often technology is expected to fix problems of undertrained staff or lack of communication. These

problems most likely will not be solved by new technology.

For example, the executive director suggests that you build a new database for tracking donors. Start by asking why she thinks the current donor package isn't adequate. She may tell you that the development director can't give her the reports she needs, because the database doesn't track that data. You see from your user inventory that the only person who received training on the database software has left the organization. After a little research, you find out that the current software is two versions behind and that an upgrade will deliver what the executive director wants at a cost of $300. In addition, the software vendor can provide training for staff to bring them up to speed. If you had not asked yourself, How else can I solve this problem? you could have embarked on a time-consuming and potentially expensive search for a new donor package, or, worse, you might have tried to develop a specialty database to do something that another software version already does. Remember to explore your options before you jump to make a change.

2. Try to build on what you have now

The goal is to have computer systems that all work together and that end users understand. Introducing a new technology will require training users on how to use it. As a result, many tech managers look to move "vertically" through an application rather than switch to a different application.

> **How do you find out what new technologies are available?**
>
> - Talk to your current vendors or consultants
> - Read computer magazines or peruse computer catalogs
> - Attend conferences or trade shows
> - Talk to colleagues at other nonprofits
> - Stay in touch with nonprofit information sites such as http://www.zerodivide.org, http://www.techsoup.org, or http://wwwnten.org.

For example, if your organization is looking for a way to track volunteers and all of your users know FileMaker, start by looking within that software for a solution, rather than by developing a different tool. (Unless after a thorough database planning process you discover a tool that makes the retraining time worth it!) Though it may seem like you are limiting possibilities, you are adapting a tool your users already understand instead of requiring them to learn a new system from scratch.

3. Talk to colleagues about how they solved a similar problem

Other organizations that do similar work can serve as excellent resources for information on particular applications or technology suited to your type of organization. Sector peers will understand the constraints you face as a nonprofit and those that affect your particular kind of organization. Find an organization that provides similar services (and, if possible, is around the same size) so you can compare notes on what tools have proven to work best for each of you. If you don't know of any similar organizations, you can look to local vendors for referrals and then call them, or post questions on a nonprofit techies e-mail discussion list. (See Techie Tools, Appendix A, for examples.)

4. Define your decision criteria

If you decide you need to purchase a new product or technology, how will you know if it will meet your needs and how will you decide between various offerings? Don't use demo packages or vendor presentations to decide. Rather, define your criteria *before* looking at any vendor's offerings so you don't fall in love with software that doesn't fit your requirements. Make sure you know what you need the product to do and how you will evaluate whether it meets your criteria. Ask the people who will be *using* the product (including frontline staff, data collectors, and report users) to help specify the selection criteria. Interview those people about how they do their jobs and how they think a system would improve their work.

When identifying assistive technology to purchase, have the user test the technology before you buy. People's needs vary greatly and while a voice recognition system might be appropriate for one user with carpal tunnel syndrome, an alternative keyboard might be better for another.

Also consider the potential risks of adopting a particular technology. Two such risks are the cost of mistaken purchases and new, unproven technology. To reduce these risks, take adequate time to evaluate the purchase— the time you invest in making the decision should be proportional to the expense of the item (and cost to replace if you make a bad choice). Work with vendors who have a known and reliable history. Make sure the technology is proven, either in the world at large or, better yet, in your sector specifically. Don't choose the first version of some software that appears to meet your needs. Buy a different program or postpone the purchase until the tool has proven to be robust.

Buy or build: The big question for nonprofit databases

Some say never build your own databases. But many nonprofits insist that nothing meets their needs—only a custom system will do. Remember: no system is going to provide EVERYTHING you want . . . particularly if you haven't figured out in advance specifically what you want a database to do! This is where database planning comes in. This is a great project for a tech consultant or volunteer, but it takes lots of input from you and your staff. Just to be clear, this planning should be done regardless of the buy-versus-build decision. Be sure to read the section on conducting a software inventory in Chapter Two and review the database planning steps and worksheets in Techie Tools, Appendix C.

Let's examine the reasons that many nonprofits give for building custom databases:

We can't find prepackaged software to track what WE do.

Even though nonprofit missions are often completely unique and perhaps the services provided to achieve that mission are also unique, the systems to *support* those services don't have to be. Ask yourself these questions if you think you can't find software that meets your database needs:

1. Are you expecting more flexibility than is actually needed? Rather than trying to cover *every* variation that happens, consider adopting a system that handles *most* types of data you collect. Occasional or nonstandard reports may need to be handled outside the automated features of the database.

2. Have you really researched enough? Many types of business software can be adapted for nonprofit uses. For example, off-the-shelf retail sales software can be adapted to track nonprofit sales. Many nonprofits also adapt QuickBooks, software that was originally written for construction companies, to manage their financial records.

We can't afford an off-the-shelf package and someone is offering to build us a great system!

All the warnings elsewhere in this book about working with technical volunteers apply here! Is this database being devel-

Finally, be sure to include compatibility in your criteria. You want to make sure that all of your tools work together.

Most final selections will be a compromise—the product will satisfy many but not all of your needs. The questions below will help ensure you pick the best compromise possible. Use these questions to develop criteria for your selection process.

- How will you use this technology? It's likely that you have an intended use for the system you're evaluating. List your answers to this question. Then describe the technology to people whom you assume will not work

oped because it's meeting a need your organization has identified, or is this particular database project driven by your volunteer's skills? In addition, ask yourself these questions about the true costs of a custom database:

1. Have you taken into account the total cost of ownership for a custom database? This includes maintenance, updating, and staff time or consultant time to develop new reports or make changes to data entry forms in a software that is unique (versus a standard package).

2. Is there adequate written documentation? Database documentation should detail how to use the database, what procedures the database tracks in the organization, and the technical information a developer needs to know to be able to maintain, repair, and update the database. This documentation takes an enormous amount of time to write and is rarely written for custom databases.

3. Who will provide user training? In addition to documentation, user training will have to be developed from scratch, as compared to packaged software for which packaged training is often available.

A custom-built database may be the answer, particularly if you only want to track a very simple set of data. But often as a custom database is being developed, or after it's been in use for a while, other functions are added. For example, an organization that has a simple Access database to track donors and donations decides to switch to a membership model. Suddenly the database needs to track expiration dates, renewal dates, and a host of other complications related to membership. What started as a simple, easy-to-maintain database now requires additional data tables, more complex data entry screens, and a system for flagging membership expiration dates. This is a good time to reexamine database needs and reconsider the prepackaged software available.

with it to see if they would use it if given the chance. A solution may be more attractive when it has features that can be used by more people in the organization.

• Does the technology address the stated need? You're not evaluating the technology because it's the latest toy; you're considering it to do a job that enables your organization to fulfill its mission. The main question here is whether it will solve key problems you listed or allow you to do something more effectively than before.

• Will the technology simplify work processes? Think about the change in behavior required to use this tool over existing methods. Will it ultimately make the users' jobs easier? Will it allow people to accomplish more or do something better?

- Does the technology improve communications or the flow of information? Will the system allow for better communication of information among staff or clients?

- Will this be useful to more than one person or group in your organization? Think about how others can use the system to become more effective in their positions.

- Is the technology easy to use? Every new tool comes with a learning curve. Spend some time using the system or seek out current users and find out how they addressed the learning hurdles.

- Can you convince people to use it? If you can determine and communicate the compelling reason that people will want to use the system, as well as provide the necessary training, you will have a higher likelihood of successful adoption.

- If we adopt this technology, what will its implications be for our other applications? Does the technology require an upgrade to your operating system, hardware, or other linked software? Have you accounted for these related costs, including installation, training, and downtime?

- What other costs might be associated with using this technology? This is sometimes referred to as "total cost of ownership" or TCO. While the software might be relatively inexpensive to purchase initially, what other expenses might be encountered in supporting the effective use of a tool? One TCO guideline is that for every $100

in your technology budget, spend no more than $30 on the equipment purchase and reserve $70 for support and maintenance. Also consider the cost of upgrading. If you have to hire an outside consultant to install or maintain the new technology, what will that cost be? Use this information when considering purchasing new technology—not just the purchase price.

- What user training will be necessary to effectively use this technology? Higher-end software requires more computer skill among users, not less.

- Is this technology accessible for people who need assistive technology? Will it work with all the existing assistive technology in place? This is an important question for existing as well as future employees. You may not have any employees or clients who need an accessible system now, but this can quickly change if someone on staff sustains an injury, develops tendonitis, or begins to have trouble seeing the monitor.

- Do you have the systems in place to manage the technology? Plan who will take ownership of the technology and ensure it is properly set up and optimized.

5. Define your decision-making process

Next you should establish a decision-making process. Your process should include such considerations as which people at your organization will be involved in the process, the timeline and budget for the whole project, and how much time and money you will invest in deciding.

6. Be sure to comparison shop

Once you've set your decision-making criteria and decision-making process, you'll be in a better position to compare one vendor's products against another.

Comparison shopping might sound like nothing more than heading over to a shopping web site like Cnet.com or PriceGrabber.com, but beware of relying on such tools as your *only* means of comparison. Search online *and* offline (bricks-and-mortar store sales), visit technology-minded deal sites like Dealnews.com, and remember that even the smallest items can have incredibly large price ranges. For example, some merchants charge two or three times more than others for an identical product such as Ethernet cable.

Purchasing Technology and Choosing Vendors

You've identified the needs of your organization for new technology. You've confirmed these needs through discussion with colleagues and an internal decision-making process. Now you are ready to proceed with the actual purchase.

The purchasing process involves more than simply comparing prices and finding the best deal. Though this process differs depending on the kind of technology that is being acquired, generally two main components of purchasing should affect your decision on how and where the purchase will be made:

1. Finding the best mix of price and vendor

2. Identifying maintenance contracts and fees

Finding the best mix of price and vendor

You've compared prices at various online and bricks-and-mortar stores, but does that mean you've identified the best price for your organization? The first rule of price-hunting for technology: *never assume that a listed price is final.*

A nonprofit organization is in a very good position to push for discounts, especially when buying items in bulk. (For example, Dell Computers has been known to drop computer prices when a multiple-unit order is placed.) Many vendors can be easily convinced to cut other deals, such as dropping shipping charges. Just remember that vendors are people too—they can be swayed by your passionate portrayal of your organization.

Many technology vendors (like vendors of almost any product) offer differing prices for identical products, depending on the customer. Let's say you need to buy a laser printer, an Ethernet hub, and a keyboard. You can purchase these online at a vendor like Zones.com or PCMall.com. However, if you had already established contact with a specific representative at one of these companies, you could have paid substantially less.

This is where vendor relations come into the picture. Establishing a personal relationship with your vendors is an excellent way to both save money and receive a higher quality of service. It comes down to the difference between how companies work with individual consumers versus businesses. Vendors will offer discounts to businesses (or organizations, in this case) because they are striving to create a pattern of consistent, future purchases (at a higher quantity than individuals will normally make). In an effort to get your ongoing business, even companies known for having strict pricing (like Apple) will offer discounts for "business" (nonindividual) customers.

You might also consider signing a multiyear contract for services like broadband or telephone service. Often if you agree to keep your service with a particular vendor, the annual price will be discounted for your loyalty. Be sure the vendor is one you will be happy with for the term of the contract. Because prices and services change so quickly, don't go overboard on your purchases. For example, many people signed expensive cell phone contracts when that technology was being adopted, only to find costs dropping radically in a few months!

Identifying Maintenance Contracts and Fees

Whether you're purchasing hardware or software, learn about the maintenance contracts and fees associated with the product in question. Some vendors may actually require that you also purchase a maintenance contract. Others might not require a contract, but will charge you exorbitant fees anytime you need phone or onsite support. Remember that the total cost of ownership for the product might be much higher if you're forced to pay a lot for service.

To avoid getting yourself (and your organization) into a sticky situation, always read the fine print and try following these general guidelines:

- Determine whether the product you're purchasing requires a contract.
- Find out all the support costs (phone, onsite, etc.). Find out if free phone support is offered, to what degree, and for how long.

 Pushing prices down

Take an hour in your day (even if you don't have immediate purchases planned) and call a vendor you often use, or, if you don't have a regular vendor, call one that has a good reputation. Ask for a small business representative and introduce yourself. Let them know that you're looking for a personal representative to work with whenever your organization needs to make technology purchases. Record all of their contact information, work schedule, and so forth. Then repeat this step with at least two other vendors. This way, each time you're making a sizeable purchase, you can e-mail all three (or more) vendors and request a quote. Oftentimes, you can then persuade them to drop their prices even more by informing them of another vendor's deal.

- Avoid long-term contracts unless the saving is significant. You always want the ability to get out of a contract, if necessary. The more flexibility for your organization, the better.

- Determine the cost of supplies for the product. For example, some printer models can require frequent replacement of expensive supplies (ink, toner, certain moving parts) and can end up costing twice the amount of another similarly priced model.

These guidelines may seem to amount to common sense, but you might be surprised at how many nonprofit organizations jump into technology purchases without much thought and end up in dire straits.

Managing Consultants and Volunteers

This next section will provide information that you can use to optimize your organization's use of technology consultants or volunteers. Keep in mind that consultants and volunteers are not employees. They do not attend staff meetings, come to the office every day, or necessarily know a lot about how your organization works. But when the relationship fit is right, consultants and volunteers can make a big difference in your organization's tech usage. It is important to manage these individuals differently than you would an employee.

Working with a technology consultant [5]

Most nonprofits use technology consultants. Often the technology being used is too complex and varied for any one person to be an expert in it all. This is why it's important to have an ongoing relationship with a tech consultant (or three!). Ask other nonprofit techies in your area whom they turn to for outside technology support, and set up interviews with some consultants. Focus on those who have experience with organizations of your size and technological complexity. We'll share more on what questions to ask a little bit later. The most important point here is to do this *before* you need help. If you wait until you're in a crisis because no one can log onto the network, you're likely to settle for the first person

It took me a while to learn that it was okay for me *not* to know everything. At my last job, I often hit the wall of my own technical limitations. I learned that calling a consultant was the right thing to do. Whenever I did that, I would observe and ask questions and note things, just to informally upgrade my skills. My boss was very flexible in bringing in outside help, and I think she respected me more after I was frank with her about my limitations and knowing when to ask for help.

— Dave Moffatt, Shepherd's Center of Winston-Salem, Winston-Salem, NC

[5] This section was adapted from curriculum designed for CompassPoint's Institute for Nonprofit Consulting, ©1997 CompassPoint Nonprofit Services.

who can show up, regardless of their "fit" for your organization and systems. Also, like other vendors, tech consultants prioritize organizations they already have a relationship with.

Accidental techies sometimes feel it's their job to deal with all things computer related. So sometimes we don't call for help when we should. Consultants can help diagnose and repair problems that take specialized skill beyond our current knowledge. Because of their expertise, they can often do it more quickly and at less expense. This runs counter to our instincts. Tackling a new problem is the accidental techie's "academy"—most of us gained the knowledge we have by diving into a problem and learning the systems along the way. But sometimes that can be costly. Tech consultants can save you many painful hours of work if brought in early on a project or purchase! Don't miss the opportunity to learn from your consultant. The time you can spend observing a consultant diagnose and correct a problem is an opportunity for you to gain knowledge and skill should a similar problem arise in the future.

What makes a good consultant?

Working with a consultant is about a relationship that depends on good communication from both parties. If the consultant doesn't tell you what he is doing in language that you understand, he might not be right for you. Be prepared to pay for the time it takes for the consultant to explain what the issue was and how he solved it—and be sure the consultant understands that such explanation is part of the job. Understanding the issue may well

give insight in how you might solve the problem yourself next time or, better yet, avoid the problem altogether. Conversely, if you can't tell the consultant what you need, you will likely not get the result you want. Here's what we think are characteristics of a good consultant.

A good consultant tries to understand the problem, not just offer solutions.

You need a consultant who asks a lot of questions to get to the root of the problem you've described. Consultants who jump to solutions

---

We regularly use an outside consultant who specializes in the technology needs of nonprofit organizations and have an excellent relationship with them. The staff knows to check with me first (using the Tech Support e-mail address). Any time an issue is beyond my expertise I get approval from the executive director, and then call the consultant.

— Barri Waye, Fairfax Court Appointed Special Advocates, Fairfax, VA

The terrible consulting experience is the guy who comes in, talks to no one as he works, gives no one any information to help themselves if it were to happen again, and disappears. We're left with a huge bill, no organizational learning, and a likely new call for the next part of the chain in a related chain of events he could have predicted and warned us about, or even prevented entirely. Grrrr!

— Gale A. Shea, Family Enhancement, Madison, WI

risk not fully understanding the problem. For example, adding a new hard disk to the server will give you more disk space (a solution), but perhaps the real problem is that no one at the organization bothers to archive files. In another year the disk will be out of space again and the problem will be back.

A good consultant learns how your organization functions, but doesn't tell you how to run it.

A consultant can be an objective, outside voice about how your organization functions. However, in providing those observations, she shouldn't be telling you how to run things. For example, a database consultant may report that the project is delayed because the development director is not completing a list of fields as expected, but she shouldn't tell you to fire the development director.

A good consultant knows his strengths and limitations.

Technology is a large field and you want a consultant who can clearly describe his strengths and limitations. For example, even though he knows technology, a network consultant isn't necessarily the best candidate to develop a web site.

A good consultant provides options, not a single package.

The value of working with a consultant is that, ideally, she is not focused on a single vendor or product but has a broad view of the options. It is also important that decisions remain in your hands based on the criteria you've defined. Technology often means making choices and compromises based largely on what is affordable.

A good consultant talks to you in a language you understand.

A challenge of working with technology consultants is that they have their own vocabulary, which can be intimidating if you don't understand what they are saying. A good consultant will help to translate what he is saying into terms you can understand.

A good consultant is willing to work with an accidental techie.

A consultant who works only to resolve the problem and won't let you observe, ask questions, and take notes might not be the right consultant for you. Yes, your organization will have to pay for the additional time spent teaching you how to resolve the problem, but, as long as the expectation of cooperation is clear to both parties, this consultant/accidental techie partnership can yield incredibly beneficial results for the organization.

A good consultant doesn't assume technology is the answer.

While you often hire a consultant to give you an answer to a technology issue, it is important that you not assume technology is always the answer. A consultant can install a great e-mail system, but that may not help staff communicate better. A good consultant will be able to identify this issue and make suggestions about nontechnical solutions.

What makes a good client?

What an organization brings to the consulting project is equally important to what a good consultant brings. In other words, you must be a good client. Here's what an organization should be prepared to offer a technology consultant to establish an effective working relationship.

A good client knows the organization's business and can educate the consultant about it.

Technology is a tool put in place to help your organization function and to better achieve its mission. A consultant needs to understand how you work in order to provide the best possible assistance. You need to be prepared to explain the work your organization does and a bit about the culture in which your work takes place.

A good client understands that organizational politics will affect the consultant's work.

If the accountant isn't speaking to the development director, the consultant will have a difficult time identifying a technology solution for exchanging information between these two individuals and, possibly, their departments. While the consultant is not an employee and doesn't need to know all your business, you need to be aware of your organizational politics and help the consultant navigate them.

A good client isn't afraid to reveal gaps in knowledge.

You need to tell the consultant when you don't understand something. Nodding your head in agreement may lead to a purchase that you didn't intend or a system you don't understand. Ask the consultant to fill in the gaps in your understanding or to find someone who can.

A good client identifies one staff person to be the consultant's liaison.

Consulting is based on a working relationship between someone in your organization—typically you, the organization's accidental techie—and the consultant. The liaison can work out conflicting requests or conflicting priorities so that conflicting messages aren't given to the consultant. An accidental techie in this role will be well positioned to easily identify repeat issues or users who could benefit from additional training.

A good client defines a job description for the consultant.

Specific projects are often defined in terms of tasks and completion dates. This works well until the project is done and the consultant relationship morphs into one of "ongoing maintenance as needed." Here the tasks are more demand driven and the completion dates are "as soon as humanly possible." You need to define the scope of tasks along with what is appropriate for the consultant to do and what the staff will do. Otherwise, you may spend more on the consultant than you need. (See the sidebar, How to Interview a New Consultant, page 70, for more advice on what to include in a consultant contract.)

A good client lets the consultant know about assistive technology needs up front.

Be sure to let the consultant know that everyone on your staff must be able to use the technology—which means the technology must be amenable to assistive technology adaptations. Don't assume that consultants will share this view or know much about assistive technology. For example, it is not uncommon for a consultant to come in and reorganize someone's system, mess with the desktop, and change the configurations and preferences. This can be a real problem if the user has set up a system so that her cursor is wider and larger, her menu bars have been enlarged, and all of her commonly used files are on the desktop in a specific location so she can find everything easily.

A good client never abdicates control of a project.

Ultimately, the consultant is working for your organization and your organization needs to maintain control of decisions made along the way. Remember, the consultant is valuable to the project for his advanced tech skill but staff are important to a successful project for their knowledge of the organization and how it functions.

A good client trains and orients staff members.

If you don't invest in training staff to properly use the tools you have, you may continually end up paying for informal training from a consultant each time someone calls. Make sure all calls to the technology consultant are routed through or okayed by the organization's selected contact person.

A good client continually evaluates the working relationship between the consultant and the organization.

Always check in with the consultant to see how things are going and to evaluate whether your needs are being met. A consultant with excellent technical skills may be a disaster if she makes disparaging remarks to staff or can't communicate competently with the variety of individuals with whom she needs to work.

What to do with your existing consultants

You likely already have consultants at your organization helping you with technology. They are part of your support structure, and one of your roles as accidental techie may be to manage requests for some of your consultants. Consultants can be an expensive resource, so good oversight can mean saving money. Here are some things to review with your existing consultants:

Meet with each consultant to discuss what he or she is doing for your organization.

Much like you created a list of technology-related tasks for yourself, you should do this with each consultant. This can be the basis for developing a mini job description for each consultant.

Ask if there are things the consultants are doing that staff could do just as easily.

Consultants often have ideas about tasks that staff can take on, but they just haven't been able to make that happen. For example, an organization asked a network consultant to come in once a week to change the tapes for the backup. At $150 an hour, that is an expensive way to back up files. Onsite staff can be easily trained to do this. Keep in mind, though, that you may need to pay the consultant to train someone how to do the task.

But it is a great opportunity to write down the steps and develop a written procedure for that task.

Ask what improvements or changes the consultants would recommend regarding your organization's technology usage.

Consultants usually have ideas about how you could improve your use of technology, but often are never asked for their opinion. Ask them.

Ask how you and your organization can better support the consultants in their work.

Consultants are often so focused on doing their task that they don't think about how the organization could use them more effectively. A consultant might be fielding calls or e-mails from a number of staff but would prefer to work through one person. Or she might find it helpful to receive a list of prioritized requests rather than have to ask each person and then decide on her own which are the most important.

▶ How to interview a new consultant

- Talk to at least three consultants.
- Ask for samples of existing work.
- Call references. Ask the references to describe the work done, their satisfaction with the work, and what advice they would give you on how best to work with the consultant if that's what you decide to do.
- If working with a consulting firm, be sure to talk to the person who will actually do the work.
- Watch for conflicts of interest.

What to ask a consultant

- Have you worked with nonprofits before? What types?
- What other projects are you currently working on?
- What role do you expect to play?
- What role do you expect my staff to play?
- How will you approach this project?
- Tell me about a project that ran into difficulties. What did you do?
- Are you familiar with access issues, guidelines, and tools for people with disabilities and functional limitations?

Hammering out a contract

- Define the scope of the project.
- Outline a list of milestones.
- Stage payments to milestones.
- Outline a communication plan.
- Discuss what to do if things go wrong.
- Define documentation requirements.
- Build in shutoff points—places where work can stop without affecting organization operations.
- Build in money for cost overruns.

Ask if you can take a few moments to acquaint the consultants with the community need and cause that your organization has taken up.

Consultants are part of your constituency and have a vested interest in your success. Keep them posted on the success of your mission and consider asking them for a donation at the end of the year.

Managing projects with your consultant

When you need help from a person outside the organization to work on a technology project, someone from within the organization must manage the project. It doesn't matter if the outside techie is a paid consultant or a volunteer, the lead responsibility for the project should stay with an employee, usually the accidental techie. Successful tech project management includes

- Setting all timelines for the project

- Checking in on a regular basis to see if targets are being reached

- Lending the outside techie assistance in obtaining information or access to resources from within the organization

- Reporting to the internal stakeholders (the executive director, board, program staff) about the project

- Confirming that the end product integrates well into your business practices

Volunteers [6]

A technology management component that is unique to nonprofit organizations is the presence of volunteers. Volunteers are everywhere in nonprofits—from board members, event staff, and folks that help with clerical duties or deliver services, to those volunteering technology services on projects like web site design and database development. This section will address the particulars of working with technology volunteers and offer some advice on how to manage these relationships.

This might be a radical statement in some nonprofit organizations: sometimes it makes sense to say, "No, thank you" to a volunteer! Because volunteers might not have the experience to understand what is really needed at your organization, the skills or time to complete the task, or the ability to communicate well with you. Even though volunteers may have the right skills, if they don't understand your organization, they may not be able to produce a product that can be used well by your staff. Work with your executive director or volunteer manager to say no to volunteers in a way that acknowledges their intent. (Example: "It means so much to us that you volunteered to help us with our database, but now that I understand the project better and your background better, we've concluded that this isn't the right way for us to make the best use of your skills.")

---

[6] Portions of this section are adapted from "Working with Technical Volunteers: A Manual for Nonprofit Organizations," ©2001 CompuMentor, http://www.compumentor.org. This excellent resource can be downloaded for free at http://www.techsoup.org/products/downloads/TechVolMan2001v1.2.pdf.

The challenge of working with volunteers, much like paid consultants, is that they aren't on staff; you, or someone in your organization, will need to manage their work and time. Also, some thought should be put into the task you invite a volunteer to work on. A project shouldn't depend upon a volunteer's skill or offer alone. For example, a nonprofit may have a board member volunteer to create a very technical and complex database. Unfortunately, no one on staff has the ability to understand how the database works or how to make design changes, nor are the skills easily learned. If the board member suddenly moves from the area and leaves the board, the organization is faced with hiring very expensive consultants to work on its "free" database.

If you do accept tech volunteers, be sure they have the skills and the time necessary to complete the entire project. And be sure you've identified the need for the project. (See the section earlier in this chapter, Assessing New Technology.) Volunteers can have many different motivations for offering to help. You'll want to clarify how much time they really have to offer and how responsible to your organization they are. Given the expense and importance of many technology projects, it's crucial to clarify what you expect from a technology volunteer. Don't skip any part of creating a volunteer job description. Think about preparing for and scrutinizing these engagements much as you would when working with any other outside consultants.

Volunteer job descriptions

One crucial step in working with technology volunteers is to create a job description. It can be much like the contract you might create with a consultant. Here are some of the items to discuss and negotiate with technology volunteers:

Tell volunteers about your organization.

Don't miss this opportunity to explain the wonderful work you do. Volunteers who understand what your organization does and are committed to your mission are more likely to fully dedicate themselves to a task. They also will be better prepared to recruit other volunteers.

PLANET 501c3 — Tales from the Nonprofit Galaxy
Unbelievable... but True!

www.planet501c3.org

JORGE FLORES OF PALO ALTO, CA VOLUNTEERED TO CREATE A DATABASE FOR AN AGENCY IN 1999 — AND COMPLETED IT!

EVEN MORE AMAZING — HE'S STILL AVAILABLE FOR SUPPORT!

REDESIGN ALL THE SCREENS? SURE, I'LL BE RIGHT OVER.

© 5/10/02 COMPASSPOINT

What makes a task a good volunteer project

- It is time limited
- It has clear outcomes
- It draws on the expertise of the volunteer
- You have time to manage the project

Examples of good volunteer projects

- Install a network
- Install or upgrade equipment
- Help select equipment
- Create a web site (as long as someone on staff is directing content and is able to maintain the site once it's created)
- Develop an information technology plan
- Assess equipment (donated or otherwise)
- Assess existing databases or software
- Offer one-time trainings

Tell volunteers how the project will affect the organization and its work.

More than simple appreciation, volunteers need to understand that your organization values their skills or knowledge.

Describe the project completely.

Be very specific about what you expect of volunteers. Explain who will be responsible for supervision. Volunteers are more invested in projects they understand and that are supported by staff!

Work with volunteers to be sure you feel confident that they have or will gain the skills to complete the tasks you have assigned.

And don't forget, you CAN fire a volunteer!

Be very clear about the timeline for the project.

You don't have the recourse you have with employees, so it is better to hand off small amounts of responsibility that can be delivered on time or reassigned. This means meet-ing with volunteers regularly once the project begins and confirming what's been done.

Discuss volunteers' commitment.

Do they want to see a large project through or just do a little piece to get you started? Is this a long-standing relationship with your organization, or is it just beginning? We've all heard stories about great new volunteers whose zeal just seems to peter out and then they disappear. Don't have a major technol-ogy project walk out the door with a volun-teer like this!

Don't forget to wrap up a project.

This is where you thank volunteers pro-fusely—perhaps have some staff take them out to lunch—and give them a chance to tell you what worked and didn't work about vol-unteering with your organization. Don't miss this valuable opportunity to get some outside perspective on your organization and how you're using your technology resources.

Managing nontech volunteers' use of technology

The other aspect of managing volunteers and technology is managing volunteers who use an organization computer to perform program work within a department. All such volunteers should be informed or trained on your organization's computer policies prior to letting them use your machines. For instance:

- Make sure volunteers follow organization protocols. If you have rules about downloading files for virus protection or are strict about maintaining consistent settings at each station, make sure volunteers are aware of them.

- Make sure a staff member is responsible for monitoring volunteers' work.

- You or the staff person supervising volunteers should train them in the computer function they are performing for you. Even though a volunteer might seem adept and comfortable with computers, it's always good to walk through the steps jointly once so you can make sure you both agree or understand what was said or needs to be done.

Assessing new technology and finding the right people to help maintain your systems is crucial to an organization utilizing technology. But how do you know that the technology you already use is as secure as possible? In the next chapter we'll discuss how you can avoid those disasters that can prevent your organization from functioning.

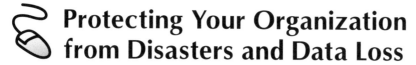

CHAPTER FIVE

Protecting Your Organization from Disasters and Data Loss

The topic covered in this chapter is the critical issue of protecting your organization from data and security disasters. Virus protection and data backup procedures are basic to an overall security policy regardless of the size of your organization or the complexity of your technology. Volumes have been written on technology security, so what follows are the rudiments on securing a network and choosing passwords. The Security Policy Checklist (Techie Tools, Appendix D) will help you plan for further protection from loss of data due to disasters, natural and otherwise. This chapter will help you make decisions on how to move ahead with virus protection and data backup.

Virus Protection

Installing virus protection software on all your computers is the *first* step. Many virus protection software packages are available.

> Talk to staff about spam and viruses. Make it funny or fun. Figure out a way for people to remember. If people understand what spam and viruses are, then they'll be less afraid of them.
>
> — Teresa Martyny,
> The Riley Center,
> San Francisco, CA

The most common solutions are installed on individual workstations. The biggest drawback to this solution is keeping each individual workstation's virus definitions updated (the part of the software that knows what is a virus and what isn't). These definitions are usually provided for a fee on an annual basis from the manufacturer of your antivirus software (Norton and McAfee are two of the most popular). As with backup systems, the best virus solutions are the ones that require the least interaction from you or your users. Many can be configured to automatically update new virus definitions on a daily or weekly basis.

After installing the software, you must be diligent in ensuring that virus definitions are updated frequently (checking every week is a good habit). If your e-mail is being hosted by an outside service, check to see what antivirus measures the host has available. If you are using Apple computers, don't gloat too much or take viruses for granted. Many viruses out there attack Apple computers, and even if you can't be infected by a Windows virus you can pass one on, possibly to your biggest funder's network along with your latest grant application.

However, all of the virus protection measures will only get you so far. Be sure to educate your users on how to manage e-mail message attachments and file downloading. Virus creators try to stay one step ahead of the virus protection definitions and are sometimes able to write code that will avoid detection by current virus definitions. For this reason users should be made aware of the basic rules of virus prevention, highlighted in the sidebar Three "Don'ts" to Prevent Virus Attacks. We suggest you copy and post this list so everyone is familiar with the rules.

Viruses can shut your organization down for days or weeks, but they can be avoided with proper and diligently updated virus protection along with well-trained and educated users. This is probably the area where accidental techies can have the greatest effect on an organization. Make sure your antivirus systems are as up-to-date and efficient as possible!

Protecting Your System from Adware and Spyware

In addition to protecting your organization from viruses, you may also want to work on controlling resource-hogging spyware or adware. *Spyware* and *adware* are programs that are installed, sometimes without a user's knowledge, often piggybacked like viruses, onto free software that the user installed. (File sharing programs, like Napster and Kazaa, are

---

 Three "don'ts" to prevent virus attacks

1. **Don't open** any files attached to an e-mail from an unknown sender.

2. **Don't open** any files attached to an e-mail unless you were expecting the e-mail—even if the e-mail appears to be from someone you know. Some viruses spread by e-mailing themselves to all of the e-mail addresses from a host's e-mail address book. Such virus-infected e-mails may appear to come from the person it stole your e-mail address from. Therefore, even if the e-mail with the attachment is from a coworker, if you weren't expecting an attachment from this person, don't open it until you have confirmation that he actually sent it and what it is. Likewise, when you send attachments, make it very clear what the attachment is and its purpose.

3. **Don't download** unnecessary files from the Internet. It's okay to download PDF files—a format used frequently for grant guidelines and publications. But program and driver files should be downloaded only by the technology staff. It's too easy to download a handy clock software utility that actually contains a spyware program.

big suppliers of spyware programs.) Spyware and adware either share advertising with the user (those annoying pop-ups or banner ads) or they collect data about how the computer is being used and send it back to the creator of the spyware. If a computer is suddenly operating very slowly, these hidden programs might be the culprits.

Free shareware programs are available to help you find and fix these issues. (See the Security section in Techie Tools, Appendix A, for web sites that offer information or links to spyware and adware scanning services.) In addition, many current antivirus software programs include spyware protection, so you should consider upgrading your software to prevent these threats.

Ongoing Software Maintenance

Most operating systems, whether Windows or Macintosh, will have updates or "patches" made available online on a consistent basis. This happens more frequently for the Windows platform than Macintosh, and is truer for newer versions of both than older (for example, the older Mac OS 9 no longer has updates released, but OS 10.x has frequent updates available). While some of these updates may be merely "suggested" and unnecessary, many resolve serious security flaws and should be downloaded and installed.

Be sure to understand the exact problem the update is repairing. Test a major update or service pack on one machine first to ensure

that it doesn't interfere with any of the programs you rely on. Take particular caution if you're running any "legacy software" (software that was designed to run on an older operating system than the current one). Make a habit of checking each workstation in your office on a weekly basis to ensure that all necessary updates have been downloaded and installed. While both the Windows and Macintosh platforms offer automatic updates, because of possible user interference, you should still confirm weekly that they have, in fact, been run.

Backup Systems

You know the drill. Without a backup system, your organization could face devastating results in the event of a computer meltdown, an officewide fire, natural disaster, or, most commonly seen, simple human error—"I accidentally deleted that file I just spent the last two weeks creating!"

In the event of a fire, flood, or any disaster that affects your entire office, there's the possibility that *all* of your organizational data could be lost at once. Every document ever created, every name and address of every funder who's ever given a grant, all of the contact information for the clients you serve—lost.

A backup allows for the replacement or reinstatement of the file or data because that data is stored someplace besides the normal location and can be restored!

The worst disaster happened as I was upgrading one of our computers. The person whose computer I was working on was gone, and I was going to copy her hard drive to a new much larger hard drive, increase her RAM, and clean up the computer. This person was one of our data collectors and was using it to produce an important annual data book. During the upgrade her original hard drive (where all her data was stored) crashed and burned. It was the first time I really heard the grinding, clicking, and end of "spinning" about which I had often read. I also found out that she had never backed up her data. This was the hardest lesson on the importance of backing up I ever experienced. All staff knew of this loss and backing up became immediately important. I learned not to trust that others have backed up before I work on their machines. The staff member did finally talk to me again—after three months.

— Lary Wells, Michigan League for Human Services, Lansing, MI

Which files should you back up?

Basically, *you must back up all of the files that you want to keep or use again*. It's that simple. All data files should be backed up—your Microsoft Word files, Excel spreadsheets, Quickbooks data files, database files, e-mail messages and addresses, and so forth.

On the other hand, you *don't* have to back up the software programs you use to access that data, like Microsoft Office or FileMaker Pro.

(Presumably you have the installation disks for those programs and could reinstall them from the CD if necessary.)

You also don't need to back up the thousands of MP3 files your coworker has stashed on his C drive, so be sure if you back up files that they are files that people actually use to get their job done! For very small organizations, this total amount of data could be as small as 100MB. Medium-sized organizations could have upwards of 70GB and so on.

Methods of backup

No universal backup system fits every organization's needs. Start by estimating the amount of data that your organization needs to back up. In doing so, be sure to allow for rapid growth in the amount of data you'll need to back up.

Once you've done that, you can move on to selecting the method of backup that is most appropriate. Many different types of backup systems are available, each with their own advantages and disadvantages. If you are overwhelmed by the idea of all your data being lost and don't know how to select a backup system, this might be a great project in which to include a professional tech consultant. Often she can help you select and install a backup system. Just be sure you understand the system and are able to test it and restore data from it!

The table on page 79 lists the most common methods of backing up.

Common methods of backup

Backup Method	Pros	Cons	Capacity
Zip disk	Inexpensive; portable	Unreliable	100MB or 250MB
CD-R/CD-RW and DVD-R/DVD-RW	Inexpensive; reliable; portable	Discs are limited in size so may require a lot of media	700 MB to 1.5 GB
Tape drive (common technologies include DAT and Travan)	Massive storage capacity; portable	Drives can be very expensive	Variable (roughly 2–500 GB)
External hard drive	Very reliable; may require more technical skill than other methods	Portability depends on price	5–500 GB
Online backups	Low startup costs; very reliable	Requires a stable Internet connection; requires sending your data to a third party; may be harder to restore data	Variable

Three important rules of backing up

Whichever method of backup you do eventually choose, you should follow three rules: rotate media and keep the copy off-site, schedule and automate, and test your backup system.

1. Rotate media and keep the copy off-site

Always keep a copy of your data off-site. This requires you to rotate between several sets of whichever type of backup media you use. We recommend you have backups of two or three weeks' worth of data, enabling you to restore pervious versions of documents if needed.

The most common off-site rotating system involves establishing a backup system with rotating tapes, discs, or hard drives. Each week, a staff member (such as you—the accidental techie) takes this tape, disc, or hard drive home or to a safe-deposit box and returns the previous one to the office.

In addition, many organizations keep a quarterly backup that doesn't get rewritten for at least a year. Be sure that the off-site storage arrangements are secure. This might mean installing a lock box in the staff person's home.

Security needs vary depending on the nature of your organization's business. If you have government contracts or have confidential client files, check to see what the contracts require or what standards your field considers appropriate for how you back up your data.

2. Schedule and automate

If backup isn't part of someone's routine, then it likely won't get done. Establish, *in writing*, a regular backup schedule and assign it to a specific staff member (or members).

Additionally, simplifying and automating as much of the backup process as possible will make the system much more consistent and reliable. Many inexpensive backup software programs can help automate the backup process and require little human interaction. The less human interaction required, the less chance of someone forgetting to do the backup. Train at least one other person besides the accidental techie so that if one of them is out of the office, staff still have the ability both to run the regular backup *and* restore data, in the case of an emergency.

3. Test your backup system to be sure the backups actually happen

Consider the possibility that when an incident does occur, either nobody knows quite *how* to restore the data or the restoration process doesn't work properly. Document the restoration process and make a habit of testing your backup system by restoring a mix of data files and graphic or text files every month or so. Restore this information to an alternate location to ensure you aren't erasing the original data if the restore fails.

Periodically review what you're backing up to ensure that any new files or folders are being included. These processes can differ depending on which backup hardware and software you use, but the important thing is to ensure that you can retrieve the data that you have backed up.

Security Systems

Gone are the days when hackers were in it just for the thrill and e-mail attachments could be opened without checking for viruses. Now large numbers of computers are infected and assembled into "bot networks," with access sold to spammers or people trying to use your network to commit fraud. Stealing personal information has become a form of organized crime, complete with software kits for setting up fake web sites to look like legitimate financial institutions. It is essential that security become an integral part of the technology infrastructure of your organization, not an add-on at some later date.

At the same time, it's important to consider the level of security that's right for your organization. Organizations that collect and store sensitive client data should be particularly aware of any federal, state, or local laws that regulate security requirements.

Wrongly implemented security measures can lock you or your users out of your systems. If you're unsure about how to implement any of these measures, you might save yourself time and invest in an outside consultant who specializes in your type of system.

Now, let's look at parts of a computer network and the security considerations for them.

Server security

Your server is a point of vulnerability and a natural target. File and database servers house much of your important (and not so important) information in one central location, and a server also "serves" by waiting for and answering requests for data.

Be sure that all security updates and patches have been applied

This means updating not only the server operating system but also the add-on web, e-mail, or ftp server, or any other applications or services running on your server.

Identify any unneeded services running on your server and stop them

If your operating system installed services that you aren't using (for example, if a web server is installed on your server and you aren't hosting a web site or administering anything through the web pages that were installed on your server), these services may present small but unnecessary risks to your security and should be stopped or uninstalled. If you aren't sure about what services are and aren't needed, check with a network consultant.

Lock it up—literally

Your server should be locked in a well-ventilated closet or private office, preferably with a dedicated electrical circuit, so that no one can physically reach it. Only a server administrator should be able to log into it.

Important risks that result from having your server out in the open include

- The janitor or a coworker could unplug your server to plug in a vacuum cleaner or a fan.
- Someone might innocently place a coffee cup or a plant above the server, resulting in an expensive liquid disaster!
- Someone could "borrow" the server's network cable to plug in their laptop.
- Someone could use the server as a workstation, and could pick up a virus or crash a program or the entire server.
- If someone can gain physical access to your server, they can gain access to the files on it.

User Workstation Security

The computers that staff use daily face a great deal of exposure to viruses, worms, and compromised web pages. Following are some important steps to take to improve workstation security.

Stop unneeded services

Just like servers, workstations often come with unneeded services installed and active. While your file and print server needs to share files and printers, your users' workstations may not need to share. If they don't need to share files or printers, turn off sharing. Look for and turn off any other unnecessary services as well.

Install a firewall at the workstation

In addition to a firewall at your Internet connection, consider using a firewall on your desktop computers. If one of your network's computers gets infected, a firewall can prevent a virus from reaching your network. Even if a virus gets through to one computer, it can be blocked by the firewalls on other computers. Depending on your network's setup and the software you are using, this can be relatively easy to do.

Consider installing and using an alternative web browser

As the market leader, Internet Explorer's vulnerabilities are frequently the target of a number of exploits. While some web sites will only work well with Internet Explorer, if your users are able to browse sites of their choosing, consider having them browse those sites only when using an alternate web browser. Many alternative browsers are available, with Firefox and Opera being two of the most common.

Network Security

Network security is important because so much of the information that you are working with travels across it, *and* because it provides a gateway to your computers and servers. To keep your data secure, you need to limit who has access to your network and what travels across it.

Be sure that operating systems, workstations, and e-mail encrypt your login information

Older operating systems and applications sometimes broadcast this information as plain text or with weak encryption.

Password protect your sensitive files

Make sure that more sensitive information (for example, financial data, HR information, client databases) is protected by secure passwords. This should be done whether you use a server or share files on a peer-to-peer network. The password should be accessible only to those who need the information.

Guard against unauthorized access to your network

Intruders can use a "sniffer" software program to gain access (for example, via a wireless connection) to your network or to any computer plugged into your network. Since a sniffer captures data sent across the network, it can access any data you send or that is sent to your computer. Given time, even encrypted data (such as your server administrator login information and other sensitive information) can be decrypted if it can be captured as it travels across the network. Active network ports (the wall jack that computers plug into) in unused offices, in unlocked wiring closets, on other floors, or in semisecured lobby areas are gateways into your network for a potential hacker, as are unsecured wireless networks. To protect your computer network from external sniffers, it is important to deactivate any active, unused ports and to secure your wireless network.

Databases should support encryption

Be sure to turn encryption on if it is available (or if it isn't available, request it from the software vendor). Encrypted e-mail connections (hosted in-house or by an e-mail provider) and secure FTP communications are also commonly available, so look for these features when considering Internet or application service providers or when setting up your own e-mail system.

Be careful when using a wireless network

An unsecured wireless network can allow someone to monitor your network traffic, steal passwords, or attempt to access your data files without having to physically enter your office. Note that even a secure wireless network isn't entirely secure and might not be appropriate for your network if your organization has highly sensitive data. If you do set up a wireless access point (WAP), consider doing some, or preferably all, of the following procedures.

1. *Change the default administrator's password on the WAP.* Be sure to write it down in case you need it later.

2. *Change your WAP's SSID (its "name") from the default to something unique.* Potential hackers know and search for WAPs that are still using the default name.

3. *Disable SSID broadcasting to prevent your WAP from sharing the name you just gave it.* Users will have to know the name in order to connect.

4. *Activate encryption, require its use, and select the highest level of encryption that it (and the equipment you want to connect with) supports.* By default, your WAP probably doesn't encrypt its transmissions. WEP (Wired Equivalent Privacy) encryption is available on most WAPs, and WPA (Wi-Fi Protected Access) encryption is increasingly available on new hardware.

5. *Update WAP and wireless card firmware.* Firmware is software that is built into your hardware, and manufactures will often issue free updates that fix security holes or that add or improve features.

If you've done all of the above but your data requires stronger security, consider also doing one or more of the following on your wireless network. These additional steps are truly effective only if you've already done the steps above.

6. *Consider using MAC address filtering.* The MAC (media access control) address is an identifier associated with wired and wireless networking hardware. Filtering will allow only MAC addresses of computers that belong on your network to communicate through the WAP.

7. *Disable DHCP on your WAP.* DHCP (Dynamic Host Configuration Protocol) allows computers that want to connect to your wireless network to request an

IP address from the router. If you disable DHCP and assign addresses to computers manually, you'll make it harder for strangers to connect to your network.

8. *Change the default network address range.* If you have disabled DHCP, hackers can still guess possible addresses to try to connect with. If you change the default address range (frequently set to 192.168.1.x) to a different private IP address range, it will be much harder to guess a valid address for your wireless network.

9. *Consider moving your WAP(s) around.* You want to provide coverage for your office but minimize the coverage that extends beyond your office walls (or ceilings and floors).

Take caution with virtual private networks

Virtual private networks (VPNs) are a way of connecting two or more networks together, usually using the Internet, so that a computer on the remote network (for example, the executive director's home computer) has access to the local network (at the office) as if it were located on the local network. VPNs can also be used to join the networks of two offices together. A VPN can bring with it special security concerns, even when properly configured, because it essentially allows users on your VPN to bypass the firewall and connect to the network. If the remote office isn't as secure as the main office and the offices are connected by a VPN, any infection or intrusion at the remote office has a back door past

your office's security and into your network. Also, think of the staff member who would like to connect from a home computer that is used by the entire family. It may not have a firewall or current virus protection and may well be infected by a number of adware or spyware applications. A VPN would plug that unprotected computer into your office network! If you do allow staff members to connect from home, consider strong policies that require them to secure their home computers the way you secure office computers—for example, require that they use a firewall, antispyware, and antivirus software, and conduct regular scans. Also it's a good idea to have remote access, enabling you to check on their protection and run scans.

At minimum, use a basic firewall

If your network is connected to the Internet, that connection allows the rest of the world a gateway into your network. You need a secure checkpoint at this gateway. Start by installing a firewall to limit the traffic in *and out* of your network.

A router connecting your network to the Internet will commonly use Network Address Translation (NAT). NAT forms a basic firewall that will block scans of computers on your network. NAT allows the router to redirect traffic from a single public IP address (provided by your Internet service provider) to and from all of the private IP addresses on your network. This limits unauthorized access to your network from the outside.

Upgrade your firewall software

A better firewall also blocks unauthorized traffic from your network out to the Internet. For example, your firewall might only allow users to check e-mail or browse web pages. If a virus or a Trojan horse infects a computer on your network and then attempts to search for other networks to infect, your firewall can block that attempt and prevent it from doing further damage.

Add hardware firewalls

Hardware firewalls are additional equipment added to your network. Hardware firewalls are generally more secure than software firewalls.

Password security

Passwords play a major role in securing your computer network. They are used to access network files, databases, and hardware configurations. If you don't choose the right ones, don't change them periodically, or fail to keep them a secret, you leave your network open to attack.

Start by selecting the right password. There are software programs available that hackers use to decipher passwords. It is easiest to decipher passwords that are short or are based on simple words in the dictionary. Many computer viruses now try a list of common passwords as they attempt to spread through a network.

 Our recommendation: Good, better, and best passwords

- *Good:* A word that is very uncommon is an okay password.

- *Better:* A word that is made up, possibly a word from your childhood. You might also combine two words that wouldn't normally go together (for example, fishhike).

- *Best:* Something that isn't a word at all.

One way of coming up with a "nonword" is to take the first (or second) letter of each word in a memorable phrase. Then add some capital letters (not just the first letter) and numbers. For example, let's pick our password as the first letters of "You want it done *when?*" That password would be "ywidw." Next, we'll add capital letters to make it "ywIDw." With this particular phrase, an easy-to-remember symbol that we can add is the question mark, making it "?ywIDw?" Because good passwords are eight or more characters in length and because we haven't used any numbers yet, let's use a number we can remember but that others wouldn't know or guess (not an address or birth date). You could choose a year when this phrase first became a significant part of your life, to make the password "?ywIDw87?" That password is nine characters long (or eleven if you include the quotation marks), includes uppercase and lowercase letters, numbers, and a symbol. Not too bad. If you use it for a while it becomes easy to remember and type in.

Don't get too comfortable with the first password you create. You might have

1. A "basic" password that you use to log into your network and perform your regular tasks

2. A "privileged" password that you use for administrator functions or when accessing restricted data that you are entitled to use

3. Different passwords for servers, for server applications, and for configuring WAPs and routers, and so forth

If this gets to be too much to remember, you might keep the passwords written down and locked in a secure place, or you can get one of the many password storing applications that require one password to get into but allow you to keep all of your passwords together in an encrypted file.

After you have selected and set all of your passwords, and you have them stored in a safe place, what do you do next? Change them! Everyone using your network should periodically change passwords to prevent unauthorized access by former employees, volunteers, consultants, or anyone else who has worked on or used your network. The less something is accessed, the less frequently it needs a new password. You might not need to change your router password for a couple of years, but your network passwords might need to be changed every six months or so, or perhaps more frequently.

Technology Policies

Technology policies outline acceptable uses, practices, and guidelines for technology in an organization. They help users understand how to keep the organization's technology secure. Just as with the finance department, technology policies and procedures affect every aspect of an organization's operation. If someone downloads a virus that infects the organization's network, preventing staff from registering clients, that breach of policy affects the organization's ability to perform its mission.

As noted earlier, when communicating technology policies to staff, explain *why* a policy is being adopted. Explain that compliance is about protecting the organization and the clients, not about limiting staff access to resources.

Like other policies in your organization, your technology policies should be reviewed by the board of directors and may be influenced by human resources policies and state and federal laws. For an example of what types of policies you might want to develop, review the sidebar Sample Information Services Policy Summary. Also visit the ENTECH web site listed in Techie Tools, Appendix A, and review the Security Policy Checklist in Appendix D.

A nonprofit techie's job can be as limitless as one lets it be. In this chapter we faced the challenging day-to-day work of systems maintenance and security. Accidental techies have so many varied responsibilities that it's important to clarify your role within your organization. The next chapter tackles that very subject.

Sample information services policy summary*

1. Do not install or remove software without written permission.

2. Hardware, software, data, and services are to be used for <organization name>-related business purposes only.

3. Do not borrow equipment for others to use.

4. Any electronic file stored on any information services (IS) system (e-mail, documents, databases, and so forth) may be viewed, moved, or deleted by authorized staff.

5. If you use a system requiring a login and/or password, you must use only the login assigned to you by IS staff.

6. Do not share your account logins or passwords with others.

7. If you bring someone (for example, a volunteer, consultant, or board member) in to use a computer or other system requiring a login and/or password for more than one day in any thirty-day period, they must have their own login.

8. You are required to log out of (or lock) your computer if you will be away from it for fifteen minutes or more.

9. Any files that you receive from outside the office network are to be scanned before use.

10. Laptops brought in to be plugged into the network must have a working, up-to-date virus scanner, and IS staff must be notified of the connection.

11. You should not use screen-saver passwords or any other system locks (software or physical) not provided by IS staff.

12. E-mail messages you send are messages coming officially from <organization name>. Use the same care and discretion as you would when using <organization name> letterhead.

* This is only a summary. You are still responsible for following all provisions of all IS policies included in the complete document.

CHAPTER SIX

Managing Your Role in the Organization

Success as an accidental techie requires more than expertise in technology systems. It also means managing your *role* in the organization so that you don't get overwhelmed, and making sure you have established appropriate support for yourself.

There are four aspects to managing your role in your nonprofit organization.

First, you must get your job in order. Review your job description and be sure your management understands your responsibilities and the activities you conduct. Because technology is one of your organization's strategic structures, your immediate supervisor as well as senior management must understand your responsibilities and activities.

Second, be sure that you've documented the processes and procedures you use.

> I decided to make it known to my supervisor that I no longer saw myself as an accidental techie, that I had decided to pursue a career in this. Eventually the organization grew to such a size that having a full-time in-house technical support person could be justified. My formal job title is now Systems & Web Administrator.
>
> — Desiree Holden,
>   Great Valley Center,
>   Modesto, CA

Third, keep your own skills and knowledge current by connecting with other nonprofit techies, either in your area or through the groups and e-mail lists included in the Accidental Techie's Resource Guide (Techie Tools, Appendix A).

And fourth, think of yourself not just as supporting others, but as a change agent in the organization for effective use of technology. Accidental techies are uniquely positioned to leverage technology to accelerate and intensify the impact of their organizations.

1. Getting Your Job in Order

As noted, to get your job in order, you need to review your job description, clarify your role with management, and document your work. We'll explore the first two of these criteria in this section. A discussion of documentation will follow in the next section.

Review your current job description

If you are like most accidental techies, your current job description doesn't reflect all that you do. By reviewing and revising your job description to include your technology duties, you will create a clearer picture of your true organizational role and the priorities of your job, and your supervisor will have a clearer understanding of the various requests tugging at you.

As you look at your job description, remember that the log of user requests you've compiled is the best place to start. This log can reveal the common technology-related activities requested of you, their priority, as well as the amount of time you spent on each task. Next, look at the tasks on your job description and hours per week that you think are needed for these tasks and their order of priority. For example, what's more important: getting out payroll or troubleshooting the network? You will need to work with your director to identify priorities and who else might be able to help with some of your tasks.

Meet with management to clarify your role

To get the truest scope of your duties, meet with your direct supervisor or executive director. Both can clarify the role *they* want you to play in the organization. Bring the list of the technology tasks you are doing. This will help to show the time required of you and help secure support for setting up the systems suggested in this book. In preparation for the meeting, you should think about some of the challenges you face and how management can help resolve them. These challenges likely include the following:

Time management

- Can I complete the technology tasks in the percentage of time allocated?
- Are there tasks that I can hand off to someone else?
- Do I have the flexibility to reorganize my time?
- Whom do talk to in order to get the flexibility?

Interruptible versus noninterruptible tasks

- What tasks are interruptible?
- What tasks require uninterrupted time blocks?
- Can I set office hours for technology issues?
- How can management help me set limits?

Coordination of roles

- If I can't attend to something, who else can do it?
- Who backs me up when I am sick or on vacation?
- How will I approach them to work with me?

Professional development

- What skills and knowledge do I need for the tasks I've identified?
- How do I learn new things: by reading, asking others, being shown how, or watching others?
- Based on my learning style, how will I develop my skills?

We're not saying that an accidental techie must become a member of the management team. However, we encourage you to formalize your role as the technology advisor in your organization and to advocate that your organization consider technology in its big-picture thinking. Taking steps to formalize your role will slowly help staff to better understand your responsibilities and theirs in complying with the systems you implement to keep your organization's technology running. Also, keep in mind that you don't have to be the only one in your organization to have responsibility for technology. A technology committee that includes staff and perhaps board members can help set priorities and begin strategic planning for technology, relieving some of the pressure on you.

2. Document Everything

To help you get your job in order, develop a job manual for yourself or a techie's operation manual. This is a guide for you and for anyone who has to fill in for you. The operations manual should contain hard copies of the inventories as well as procedures for which the accidental techie is responsible: how to back up the server files, how to update virus definitions, how to create a new user account, and how to set up e-mail access for a new employee. In addition, the organization's vendor or tech support contacts should be listed.

A good way to think about what to put in this manual is to think about anything you do daily, weekly, or monthly, and include instructions about how to do it. The manual should provide someone coming in to manage your technology with a good picture of what needs to be done. Wouldn't it have been nice to have such a manual when you got this job? If you need another reason to create this manual, ask yourself this question: Do you want to take a vacation some day?

Following are some procedures that could be documented in a technology operations manual:

- Updating virus definitions
- Backing up and restoring files from the server
- Updating extensions in the phone system
- Setting up and using e-mail
- Who to contact when the e-mail system fails
- Rebooting the server
- Setting up new employees on the network and giving them an e-mail account
- Setting up reports and layouts for your databases
- Changing web content and testing for disability access when changes are made to the web site

> Document *everything* you do. Even though there isn't enough time, the organization will be up the creek without a paddle if for some reason you have no documentation and unexpectedly have to leave. If you document along the way instead of trying to document everything at once, it doesn't take as much time.
>
> — Teresa Martyny, The Riley Center, San Francisco, CA

How to document procedures

When documenting procedures, write down each step as you perform the task. Assume the reader will have *no* knowledge of a task except the instructions that you provide. Record each keystroke, mouse action, or other piece of information needed. To maintain security, do *not* include passwords. Include screen shots if possible. Here are questions to consider as you go about creating the documentation.

- **What is the procedure?** What exactly are you trying to accomplish with this procedure? This might be obvious to you, but don't assume it will be to the person attempting to accomplish a task.

- **Who is doing this now or who knows how to do this?** It's good to know who is presently responsible for the task. They should help document the procedure. They might have a shortcut or know something that hasn't been previously documented.

- **Who is primarily responsible for doing this procedure?** Whose job will it be after the documentation is complete?

- **Who is responsible for backing up this procedure?** If the person responsible is unavailable, whom do people go to for help with this procedure?

- **How often is this procedure performed?** Is this something that needs to be done every day, every week, or once a year?

- **Why is this procedure performed?** People are much less resistant to following procedures when they understand the reasoning. To prevent appearing punitive, explain *why* users are being asked to do something. If users are being asked to log off of their workstations when they are away from their desks, explain the security reasons for logging off. If you're asking people to do more work, or different work, they need to know why!

- **Where is the equipment located?** If the instructions are to perform a procedure on a particular piece of equipment, tell the precise location of the equipment. For example: "The workstation labeled MAIN-SERVER that is located in the server closet to the left of the door."

3. Connect with Other Nonprofit Techies

One of the most useful things you can do is find other techies in organizations similar to yours. We recognize that this might be easier in some parts of the country than others, but if there's not a group near you, consider starting one or consider going to one of the N-TEN (Nonprofit Technology Enterprise Network) regional conferences. You could also attend one of their 501 Tech Clubs if your city has one. Join the Techies e-mail list or log onto TechSoup. (All of these resources are listed in Techie Tools, Appendix A.) The important thing to know is that you are not the only one out there doing what you're doing! Connecting with other techies, particularly those who are also working in nonprofits, is a great way to learn some tech tricks, and meet lots of helpful experts who are often willing to answer your questions.

4. Become a Change Agent within Your Organization

Getting your job, documentation, and peer connections in order is the start, but your real goal is to ensure that technology is used effectively in service of your organization's mission. This is the art and joy of an accidental techie's life: identifying opportunities for change and growth, and shepherding them into place. Here's a four-step process that may be helpful.

Step One: Identify the change

Identify what technology change needs to happen. The technology change doesn't have to be the *absolute* solution to a problem, but it does need to make the delivery of services more effective or make people's jobs easier. Use information from your inventories and tech support request logs to determine what your organization's technology goals and needs are. Conduct interviews with key technology users and decision makers at your organization to get their perspectives on how well technology is working. You will gain a sense of what your organization's management knows about technology and what they think would improve overall operations.

Step Two: Determine where in the organization change can best happen

After you've decided *what* your organization's technology change goals are, decide *where* the change you seek needs to occur. Is the change needed in some procedural or mechanical area? For example, is the problem in a system,

structure, or piece of equipment that needs to be replaced or updated, or is the problem caused by inadequate training of staff? If your procedures and mechanics are correct, you must determine whether some aspect of your organizational culture is at the root of the problem. Is there a fear of technology or a fear that technology costs too much? Is there an inability to see the change strategically, which leaves people unable to see the benefit of the proposed change?

Step Three: Determine who has the decision-making authority

Once you know *where* change should take place, you have to find out *who* can make the change happen. Is it the board, the executive director, a program manager, the staff, or someone else? Once you know whom you have to influence, figure out how to influence them. Again, you may be attempting to influence management systems that you may not have been formally invited into. So the question becomes how to manage technology when you are not considered "management."

As we've mentioned, one of the best ways you can influence the organization is to formalize your role as the person responsible for technology in the organization. Another very helpful approach is to answer the number-one question in strategy setting: "What's in it for *me*?" That is, clearly identify the advantages of the change you'd like to see from the perspective of the person who has the decision-making responsibility. Tell the executive director how it will make the organization more effective. Tell the clinic manager how it will make her employees more efficient. Tell frontline staff how it will make their job easier. You need to know what is important to whomever you are trying to influence and then articulate how the change will benefit them.

Step Four: Develop a strategy

After defining what is needed (Step one), where it is needed (Step two), and whose support is needed and what you need to communicate to gain that support (Step three), it's time to figure out how to move forward. The method you use to develop your strategy may be as unique as your organization and the tech need you have identified. Your strategy may be developed as part of a strategic planning or budgeting process. It may be developed from the one-on-one discussions you've conducted or as a result of the last tech crisis you faced.

Examples of the accidental techie as a change agent

Following are some examples to inspire you as you help your organization adopt new technology habits. Each reveals how you can make change happen, regardless of your authority in the organization.

Influence through others

An administrative assistant (and accidental techie) at a small HIV prevention organization was frustrated with the e-mail system. Seeing the potential to use e-mail to communicate with clients, she pitched her ideas to the client services manager. The manager agreed that programmatic improvements could be achieved if more people used e-mail.

Fortunately, the organization was in the process of creating a strategic plan. So, the administrative assistant/accidental techie asked the client services manager to help influence the executive director and board at the next board meeting by explaining how the change would make the HIV client services department more efficient. The timing was excellent, because the board and executive director were looking for ways to improve mission-related productivity. With more information at hand to make a decision, the decision makers were better able to see how the desired technology change could be incorporated into the strategic plan.

Identify the *real* decision makers

The executive director of XYZ Family Service Organization wanted to switch the organization's data collection from paper-based systems to an electronic data collection system. However, many of the staff had worked at the organization for years and valued the paper-based systems. They were concerned about privacy of electronic data collection, not to mention their discomfort with learning a database. In addition, the organization's board of directors was very concerned about the cost of changing to an electronic database.

Even though the executive in this organization had a lot of power, resistance from staff and the board made *them* the real decision makers. (For change agents, decision makers can be thought of as *anyone* who has to be influenced to make the desired change happen.) The staff had to be influenced to participate in documenting the current collection systems and convinced that electronic collection would be secure *and* make them more efficient when serving clients. The board had to be convinced that the financial risks to the organization would be minimal.

To quell the board's fear and engage staff in the process, the executive director allied with two other organizations that did work similar to XYZ. The three organizations together developed a detailed list of needs and requirements and researched available off-the-shelf databases. The fact that the board members of the allied organizations thought the undertaking was important reassured the XYZ board members. Plus, the risk (staff time) was spread out among all the organizations. Because of their involvement in the database selection process, XYZ's staff was more open to learning the new database and using their new desktop computers. They felt good about the organization's commitment to the database project.

Show decision makers that the change will work

Our third change agent is a board member at a small organization with an active, task-oriented board. This board member *really* wanted the organization to have a web site. Other board members and the executive were uninterested. The board member, a hands-on techie, went ahead and created a small demonstration web site to show what could be done. While showing off the site at a board meeting, the board member committed to working with the staff and board to fully develop and test a site that would serve the organization's mission. By minimizing the perceived risks of the initial web design process—and by showing the real benefits—this board member was able to illustrate the web site's potential without investing staff time or hiring a web site consultant.

Many accidental techies ask, Why do I have to convince decision makers? Why don't people just get it? People have different ideas about how or whether things should be changed, because their knowledge and experience differ. When considering technology change specifically, many people are uncomfortable with technology and resist it. Because of that, many decision makers will be slower to accept technology change. As a change agent, you can succeed by identifying the real decision makers (or decision blockers), enlisting the support of others to influence them, finding ways to reduce their resistance, and offering concrete evidence that the change will produce real benefits. Each of the examples above contains elements of these factors.

Some last pointers for your role as change agent: Your organization probably can't address every opportunity you identify, so focus on the changes that will benefit your organization the most. Also, find an ally whenever possible. A board member or manager with a good understanding of technology can be very helpful in moving your technology strategy along in the organization.

In this chapter we advocated that accidental techies become technology managers and promoters of effective technology change within their organizations. We also looked at ways to influence the technology decision-making process. But even if everyone is on board with a proposed technology change, that change usually costs money. In the next chapter we'll get some advice from Eugene Chan, director of technology at the Community Technology Foundation of California, about some of the strategies accidental techies can use to pursue funding for technology at their organization.

CHAPTER SEVEN

Finding Funding for Technology

Working as an accidental techie, you know it takes more than elbow grease and user manuals to keep computers and networks running—it helps to have money to support the technology. Likewise, finding funding for technology also requires more than elbow grease and letters of inquiry.

The article in this chapter, written by Eugene Chan, director of technology at the Community Technology Foundation of California, focuses on how accidental techies can help the fundraising prospects for an organization. Don't worry, we're not adding "accidental grantwriter" to your list of innumerable duties.[7] Just keep in mind that fundraising is a team sport and you are a vital part of the team. As the resident techie, you are in the best position to articulate the ways technology touches the operations of your nonprofit, and to ask funders for the right type of resources for your technology plans.

> I am actively involved in the technology decisions that are made at my organization. I am on the fundraising team and specifically try to find technology funding, although it has been tough. Most groups want to fund the service that you provide to the community and not the technology that allows you to provide that service.
>
> — Stacy Smith, Housing for Mesa, Mesa, AZ

For funders who read this chapter, we want you to see the view from a technology grantseeker's perspective. Whether or not you identify as a "technology funder," you will likely receive requests that employ technology as an element or strategy within a proposal. And if you do prioritize technology, how can you make your funding more accessible to organizations that rely on homegrown technology strategies such as accidental techies?

Whether you are an accidental techie, an executive director, or a funder "eavesdropping" on this chapter, here are three things you need to know about

- How technology is funded
- How funders assess technology requests
- Six ways nonprofits can leverage technology investments

---

[7] Unless, of course, your official job responsibilities truly include *grantwriter*. In our experience, accidental techies often emerge from fundraising or finance departments by virtue of their desire to ensure that donor tracking or accounting software keeps running.

How Technology Is Funded: The Basics [8]

Talking with more than one funder about technology is akin to asking a roomful of people to look through a kaleidoscope pointed out a window. Even with the same instrument and the same view, each person will likely describe a different scene before them.

Consider this: a foundation recently awarded a technology grant to a nonprofit for a network server, several personal computer workstations, and outside consulting assistance. What do you think is the overarching purpose of the grant?

A. To replace an unstable peer-to-peer network and to enable a new fundraising database to be installed

B. To close the digital divide between technology haves and have-nots in the community

C. To help local youth develop new digital media and content as a form of self-expression

D. To access a funding organization's intranet as a tool for reporting and collaboration with other nonprofits

E. To create a new model of community organizing and advocacy via the Internet

The correct response, as you may have already surmised, is that each answer represents a strong and compelling, yet different, reason

why a funder would support a nonprofit's technology. Funders arrive at technology funding from different places—some view it as essential capacity building, others as a way to facilitate communication, connection, and collaboration, while still others see it as a key component to empower and engage clients through community-based services and programs.

What funders of technology do have in common is a desire to see that the implementation of technology is successfully aligned with the mission of recipient organizations. In your quest to raise resources and obtain grants, tap into the funder desire to ensure that, above all, your organization's ability to meet its mission is well served by the technology you are seeking.

Cash grants

In 2002, according to the Foundation Center, U.S. foundations awarded 1,856 grants representing $157 million worth of support toward computer systems and technology to nonprofits.[9] As sizeable as this figure may appear, it was only a 1 percent slice of the philanthropic funding pie for that year.

As a grantseeker, following universally good grantseeking practices will do more to build positive relationships and credibility than your ability to deploy the latest version of a database or to relaunch your web site with a content management system. When it comes

[8] By Eugene Chan.

[9] "Types of Support Awarded Foundations, circa 2002." *Foundation Giving Trends*, The Foundation Center (2004). Available at http://www.fdncenter.org/fc_stats/pdf/07_fund_tos/2002/15_02.pdf.

to technology grantmaking, funders often say "never lead with the technology." But at the same time, it is the appropriate application of technology that will determine the successful implementation of a grant. It is likely that technology is on the radar of every funder who is concerned with capacity building, data collection and measurement, communications and outreach, disability access and cultural competency, electronic media, skills training, culture and arts, not to mention the direct impact of technology in communities themselves.

As an accidental techie, what can you do to help improve the prospects that your organization will get funding for needed technology?

- **Learn the basics of grantseeking.** If you are not a grantwriter, make an effort to learn the best practices in resource development for your nonprofit and what makes for quality proposals. Resource centers all across the country carry information about foundations including copies of annual reports, funding guidelines, searchable databases, and grantwriting workshops. For example, the Foundation Center has libraries in New York, Atlanta, San Francisco, Cleveland, and Washington, DC, that you can visit, along with an informative and in-depth web site for online visitors (http://www.foundationcenter.org).

- **Document your organization's technology assets and stories.** By this we don't mean just providing an inventory of computers and equipment. We refer to being the keeper of institutional knowledge at the nonprofit to identify instances when technology proved its strategic worth. It might have been in the development of an important database for a membership campaign, or the quick launch of a web site that educated visitors about an impending public health crisis, for example. You are in the best position to articulate why and how technology has made a true difference in the impact of your organization.[10]

- **Provide input and information on technology-related portion of budgets.** This is important at both an organizational level (for operating budgets) and at a programmatic level (for proposal budgets). Program officers, especially those whose expertise is not in technology, often consult with their information technology staff when reviewing technology proposals. In addition, seek to include technology overhead into every funding request, regardless of whether the proposal itself is technology oriented.

- **Be a technology "translator."** Funders, nonprofits, and techies often communicate in what seems like wildly different languages. This often happens in person (at meetings) and on paper (in proposals). As an accidental techie, strive to communicate

---

[10] See Chapter Six for suggestions on documenting your work and, in particular, ways you can be a technology change agent within your nonprofit organization.

information about technology clearly and in plain language.

- **Carefully and regularly appraise your technology costs and consultant bids.** What is the difference between a $1,000, $10,000, or, even, $100,000 web site? Prior to writing it into a proposal, examine the technology costs for bids and ask potential contractors or vendors to support their numbers instead of simply "cutting and pasting" them into the document. Remember that a program officer will look closely at your numbers—even if you haven't.

- **Talk with other accidental techies about possible funding sources and opportunities.** Especially if you have targeted a specific funder for a request, do additional homework by talking with your counterpart at another nonprofit that has already received a grant from that funder. Ask if they would be willing to make an introduction or provide a reference for you. Build relationships with program officers who oversee grant portfolios whose priorities match those of your organization. Learn about funders, their priorities, and their patterns of giving by reading annual reports, web sites, and grant guidelines.

In-kind donations

Recognize that technology funding doesn't always mean cash grants. In-kind donations of hardware, software, or services should be a valuable part of your resource development strategy and may be easier to obtain than cash grants, especially from technology companies.

If you have been granted a technology donation, be sure that it is compatible with your existing technology and systems. (Your technology inventory and assessments can guide you toward what types of equipment and software will be acceptable.)

And don't limit yourself to just thinking about compatibility at the equipment or application level. New hardware or software invariably means new training and support needs for you and the rest of the staff. Graduating users from one version of an application to the newer version will require less retraining than migrating to a different piece of software altogether.

If you are preparing to seek donations from companies, also inquire whether they have employees, especially techies, who would be interested in volunteering with your organization. It may be possible to get discounted or donated services from Internet service providers (web hosting) or application service providers.

In the end, treat in-kind donations as valuably as you would a million-dollar grant. Acknowledge the gifts publicly and privately. Be a role model for how a nonprofit can creatively apply technology to your mission. Keep the donors abreast of how you are using the technology and the impact it has made on your organization.

How Funders Assess Technology Requests

Funders want to help—but they also want to ensure that their investment is going to be used effectively. Funders want you to demonstrate sufficient technology capacity and a sustainability plan for technology. Most funders consider at least the following five questions as they assess whether to support technology requests:

- Is it too expensive?
- Is it too complicated?
- Is it too much to do at once?
- Does it relate to the organization's mission?
- Does it relate to the foundation's mission?

Imagine yourself at a site visit where a local nonprofit is seeking funding from a foundation to upgrade its information technology and client tracking system. Let's see how each of these questions applies.

Is it too expensive?

FUNDER: The price tag of the technology purchase coupled with the ongoing costs of technology maintenance seems steep.

NONPROFIT: While the budget may appear high, for this technology project we looked at several alternatives. We determined that this current configuration maximizes the impact that the technology will have on our organization and will be cost-effective in replacing several separate pieces of equipment. We joined with two other organizations that are also upgrading their technology and have negotiated an additional discount from the vendor. Some of the designated software was available through TechSoup Stock for only an administrative fee. All in all, we've been able

to lower the effective cost of the project, and still have the same level of technology functionality.

Budgeting for the cost and the life cycle of technology is a key part of being an effective technology grantseeker. Drafting a technology plan is a good way to begin seeing an organization-wide picture of your technology needs and costs. It also allows you to plan beyond one single grant proposal or one single funding source for technology.

Is it too complicated?

FUNDER: How do you plan to manage the installation and the maintenance, especially since you only have an accidental techie for the support?

NONPROFIT: It's true that our accidental techie is a generalist, but she built the original network and bought the first batch of computers. She is the person most familiar with the technology aptitude of our staff. This also builds on her skills and experience with previous versions of this brand of equipment. Through our recently drafted technology plan, we have analyzed our resource needs beyond the initial purchase and will build into our operating budget upgrades to the technology over the next three years.

If you can, propose the simplest system that addresses your requirements. Once again, a technology plan should aid in simplifying a road map for your organization.

Is it too much to do at once?

FUNDER: How realistic is it that you will be able to implement everything in this proposal while still providing for everyday services?

NONPROFIT: We've planned to roll it out in two phases over the year, but we'll still need to push hard to hit the timelines and milestones we listed. We checked in with other organizations that have launched a similar project. They told us that having an adequate and secure physical space was one of the critical paths, which we've already completed. We also hired an outside MIS consultant, recommended through a techie's listserv, to help with the initial installation, more complicated troubleshooting, and the training of our accidental techie.

Three key resources when it comes to technology deployment are *money, time,* and *staff.* While you may not have full control over what types of money will come to your organization, you do have better ability to control the timing and the timeline of technology projects.

Does it relate to the organization's mission?

FUNDER: Launching this technology is a new area for you. How does the technology change what you do?

NONPROFIT: It's very much in keeping with the mission of the organization. Our client population, which was once located only in this neighborhood, has slowly spread out regionally. We see the technology as giving us the tools to better stay in touch with the clients.

We'll also be able to better coordinate services by publicizing events through our web site and tracking data in our case management system. Put simply, the new technology will allow us to do what we do better and represents an extension of our service area.

Often, a proposal will contain technology support that seems new, but actually supports tools or projects that an organization has worked on for a period of time already. Or, if technology has not been a core area of expertise or part of your programming, consider partnering with another nonprofit for the funding request. Either way, ensure that the link between the proposed technology and your organizational mission is clear and direct.

Does it relate to the foundation's mission?

FUNDER: You haven't explained the technology well enough for me to make the case to fund it.

NONPROFIT: As an existing funder of ours, you know that we share your goals of promoting educational and economic opportunity in this community. The proposed funding will do three things to improve our current technology infrastructure: (1) staff time will be saved, especially for our accidental techie, because we will spend less time fixing unexplained crashes; (2) it will allow us to access and share information electronically so that our service area and time in the field can be greater; (3) with the new database management system, we will be able to better track and report our outcomes to more accurately reflect the impact of our program services in the community.

The bottom line is that funders support programs that meet your mission and theirs. They are concerned with effectiveness, impact, and scale. Many funders are also worried that technology proposals designed to increase access and reduce barriers wind up making things worse without adequate planning. Demonstrate how your plans for technology and the role of the accidental techie align to help you meet the goals of your organization and your funder.

Six Ways Nonprofits Can Leverage Technology Investments

As you think about how to grow technology capacity and recognize that increased fundraising will be a key part of that growth, think creatively about ways to leverage resources you already have at hand to lay the foundation for effective and compelling technology requests to funders. Here are six different ideas nonprofits can use as starting points.

1. Be your own lead technology funder

Often, foundations may not be ready to be the lead or sole supporter of a technology project. But without that first funding partner, how do you establish a track record and credibility?

If your organization has the financial wherewithal, consider devoting a onetime portion of your budget to a technology project for your organization. In short, make a technology grant to yourself. Here are the steps:

A. Outline the problem and state why technology makes sense as a solution

B. Plan and develop a proposal (including a realistic budget and timeline)

C. Review the proposal and challenge its assumptions (if you can, ask a nontech staff person, a board member, or even a current funder to help in this review)

D. Make the decision to fund (or not) and set measurable objectives for evaluating success

If you can say that you invested in your own technology in this manner with deliberation and forethought, funders are more likely to be convinced to commit resources to the effort. And undergoing such a process may enable you to better anticipate issues of concern that a program officer may raise when you submit the actual proposal for funding.

2. Support training for accidental techies

Think about contacting funders for resources to provide classes, training stipends, or scholarships for accidental techies. Accidental techies rarely have the opportunity to receive formalized training in help desk support, database management, network administration, or in any of the areas outlined in this book.

These days numerous learning options (in-person and online) are available, especially in the information technology arena. Classes tackling every conceivable aspect of technology are offered through local colleges or specialized for-profit and nonprofit training providers. Attending technology conferences is a great way for techies to compress learning into a short period of time and to meet other techies as well. Online classes, or e-learning, are an option that avoids the travel and time constraints of classroom-based learning.

Given the quickly changing nature of technology, investing in the skills of accidental techies can be a low-cost investment that yields a high return. With this strategy, you may be able to tap into "education"-focused funders, rather than just "technology" funders.

3. Map critical technology functions to program functions

Examine your organization's critical administrative and program functions and how technology helps you to carry out these functions. You can do this simply in a word processor as a table, or, for the more visually oriented, it could be graphically mapped out in a diagram. We suggest this as a good exercise to do during a staff retreat or strategic planning session where the entire staff can participate in mapping out these technology linkages.

For example, such a map may lay out the need that communications (which includes outreach and development functions) has for a robust contact database, individual e-mail addresses for staff, and a regularly updated web site. The map would outline who is responsible for maintaining the applications, the content, and the required technology each person needs to perform their responsibilities.

Not only is this a good strategic planning tool for management, it serves as a handy way for organizations to document to funders just how integrated the technology functions are in your nonprofit's operations and programs. It can also be the precursor to a full-fledged technology plan. Just as organizational charts can provide clarity between staff positions and program functions, a technology function map will serve to explicitly link your information technology infrastructure to your nonprofit's organizational capacity.

4. Use nonprofit technology assistance providers

If your organization has utilized the products, services, or assistance of technology-focused nonprofit technical assistance providers (NTAPs)—CompuMentor, NPower, Nonprofit Technology Enterprise Network (N-TEN), and CompassPoint Nonprofit Services come to mind—you have reaped, in essence, the benefits of grants made by funders who care about technology and nonprofits.

Visit online technology providers such as TechSoup (http://www.techsoup.org) or Gifts in Kind (http://www.giftsinkind.org) for nonprofit-priced technology. Collectively, NTAPs are an effective way to contract for technology consulting, obtain quality training, and acquire affordable technology (especially software).

In addition to NTAPs, explore local alternative technology support resources, such as established community technology centers, circuit rider programs, or college internship programs, to augment the technology ability of your organization.

5. Understand the total cost of ownership of technology

Say you are shopping for a new car. You've bought and read automobile magazines, read online reviews, and visited dealers. You know exactly how much you can spend. Beyond the sticker price, an informed car buyer will also weigh factors such as fuel mileage, insurance premiums, and car reliability when deciding whether a car is affordable. Wouldn't you? Now, what if we applied this thinking to technology investment as well?

This is what *total cost of ownership* (TCO) does and why it can serve as a useful guide for nonprofits and funders alike. A concept pioneered in the corporate sector, TCO refers to anticipating all costs associated with the deployment of technology: equipment, software, ongoing maintenance, troubleshooting, and even staff downtime resulting from technology failures.

A suggested TCO guideline is that for every $100 in your technology budget, spend no more than $30 on the equipment purchase and reserve $70 for support and maintenance. More abstractly, anticipate that 70 percent of the technology budget will go to the costs associated with ongoing maintenance, upgrades, unanticipated downtime, and troubleshooting of technology.

For nonprofits, applying a TCO rule of thumb enables you to plan for growth of technology support resources that is consistent with your organization's growth. For technology funders, understanding the TCO of a project is an important measure of understanding how a nonprofit has planned for the sustainability of technology after the purchase has been made.

6. Share knowledge and resources

While each nonprofit has a unique mission and constituency, the experiences and challenges that face accidental techies lie along roads well traveled by other individuals and organizations. Consult with other nonprofits to see whether they would be willing to exchange copies of technology plans or funding proposals with you. Or check with a local nonprofit technology assistance provider who might have sample templates or worksheets.

I first learned about HTML and web pages from a past employee who had worked as a web projects manager before she came to the Great Valley Center. From there I began to do small projects like creating photo albums for past events or workshops. Over time I gained the organizational skills needed to maintain the entire web site.

— Desiree Holden, Great Valley Center, Modesto, CA

Without copying, seek to learn and benefit from other nonprofits' insight, planning, and approach to technology. Actively ask and answer questions. Adapt and apply (with permission and acknowledgment) what you can to your own organization and technology systems. And, if you ever find yourself with an opportunity to share information or resources that would benefit another organization, especially in the realm of technology, do so with good cheer and a helping hand.

This chapter outlined the important roles that accidental techies can play in resource development and fundraising, such as helping to document the value of technology and to prepare of technology-related proposal plans and budgets. Because technology funding is still a relatively small portion of overall giving by foundations, apply all the tools and evidence you can to make the case to your current and potential funders for the effectiveness of a technology investment in your organization.

Don't forget to be an ambassador for the value of technology resources and investment. Whether the end result is better communications, improved data analysis, or a more dynamic web site, if you believe in the promise and potential of technology, evangelize ways that grantmakers and grantseekers can reap the benefits of technology.

Technology and the Work of Nonprofits

Technology has become an integral part of the operation of most nonprofit organizations. The need for a person responsible for managing what is often a major portion of the organization's budget and mission-critical functions has placed many people in the role of accidental techie. A formal technology role, filled by a person who has the authority to manage, can greatly affect an organization's performance.

As we've discussed, the accidental techie can have any number of jobs at the organization. The techie can work with outside consultants and be the in-house tech wizard. Responsibility can be limited or varied. Regardless, the most important things that an accidental techie offers are knowledge of the organization's mission-driven work; an understanding of the personalities, skills, and potential of the people carrying out this work; and the means by which technology can be applied to further the services the organization provides to clients.

We hope that accidental techies take away from this book a better understanding of the complexity of the job they do. The combination of work that requires very focused concentration (like database design or web design) with tasks that require immediate attention (like getting the network back in service) would overtax a Super Techie—so think of the impact on an average techie. As we've seen, it's not just the continual acquisition of technical knowledge that makes the techie's job so tough. It's the nature of technology use and management in most nonprofit organizations—almost as though it's an afterthought—that can make the job so stressful.

We hope that nonprofit leaders and directors now have a better understanding of the techie role. We hope these leaders see the importance of supporting a more formalized technology practice and including technology as part of all planning and budgeting processes.

We hope that funders also now have a better understanding of the challenges nonprofit organizations face and how important it is for them to consider technology when designing capacity building or strategic planning projects.

Whatever your role at a nonprofit organization, you should now have a clearer picture of how an accidental techie contributes to the hard work nonprofits do to provide service, change lives, and obtain social justice for all—and how we can best use technology to do that work!

APPENDIXES

Techie Tools

The appendixes offer great tools to help you in many of the projects this book describes. You'll find blank forms that you can use immediately in your organization, more details on certain processes and terms, and many links and resources where you can find information for your technology projects.

Be sure to visit **Accidentaltechies.org**, the companion web site to this guide. You'll find electronic versions of the forms, tools, and other resources. The site will be updated periodically to ensure you have access to the most up-to-date information.

The Accidental Techie's Resource Guide

Note: All sites were active upon publication, but as an accidental techie, you know that web sites change as fast as the weather. If you can't find a listed site, search on the name—and please notify the publisher. Thanks!

General and nonprofit specific technology advice

Accidental Techies Web Site

http://www.accidentaltechies.org

This web site is the online companion piece to this book. All of the links and worksheets in the appendixes are included on this site.

Cnet

http://www.cnet.com

Cnet features reviews, articles, and how-to information about technology (similar to ZDNet; see listing at the end of this resource section).

501 Tech Clubs

http://www.nten.org/501techclub

501 tech clubs are informal groups that meet monthly in a number of cities so that people helping nonprofits make more effective use of technology can meet their local colleagues, develop a professional support network, and "talk shop" in a relaxed setting. The name reflects the fact that group members' clients and employers are primarily 501(c)(3) non-profit organizations.

Global eRiders—Mission Driven Technology Support for Nongovernmental Organizations

http://www.eriders.net/

eRiders are roving technology consultants who work one-on-one with a group of related nongovernmental organizations (NGOs), helping each organization to develop and implement an information and communications technology strategy tailored to its unique aims, needs, and context.

Glossary of Computer Terms

http://www.ugeek.com/glossary

Where to go when somebody tells you that your IRQs are conflicting on the PCI bus and you have NC (no clue) what that means.

Information Technology Professional's Resource Center

http://www.itprc.com

A good source of general information for IT professionals.

IT Resource Center

http://www.itresourcecenter.org/index.shtm

The IT Resource Center helps nonprofit organizations use computers. Founded in 1984, the nonprofit center provides comprehensive technology planning, training, and support services to more than four hundred organizations each year.

The LINC Project

http://www.lincproject.org/

The Low Income Networking and Communications (LINC) Project is an undertaking of the Welfare Law Center. Since 1998, the LINC Project has worked to build the technology capacity of low-income grassroots groups across the country, enabling community-based organizers and low-income individuals to gain a presence in public debates on economic justice issues, become informed, reach new allies, educate communities, share strategies, and participate in the democratic process.

NPower

http://www.npower.org

NPower is a network of independent, locally based nonprofits providing technology help that strengthens the work of other nonprofits. Check out the tools, which include TOC and ROI calculators.

N-TEN (Nonprofit Technology Enterprise Network)

http://www.nten.org

N-TEN helps nonprofits understand and employ technology effectively. It creates opportunities for nonprofit staff and technology support providers to identify peers and develop professional support networks, share information and resources, and work collaboratively.

One/Northwest

http://www.onenw.org/

Technology assistance for the Northwest environmental movement.

Tech News

http://www.uwnyc.org/technews/

Tech News is published online for human service agencies and other interested organizations by United Way of New York City with support from the IBM Corporation.

TechRepublic

http://www.techrepublic.com

TechRepublic is a good source of information for people working in information technology. You must sign up for a free membership to get full access. The site features many downloadable word documents that you can adjust to use as your own forms and has articles on many areas of technology management and support.

TechSoup

http://www.techsoup.org

TechSoup is set up to be the portal to the nonprofit technology world. It also contains links to TechSoup Stock, formerly known as Discount Tech.

TechSurveyor

http://www.techsurveyor.org

TechSurveyor was specifically developed to help nonprofits inventory and manage their technology. By using TechSurveyor with a TechAtlas account, nonprofits can keep track of assets and technology planning with one account for their organization.

ZDNet

http://www.zdnet.com

ZDNet is one of the largest sites for reviews, articles, and how-to information about technology. It publishes many major computer magazines such as *Computer Shopper* and *PC Magazine*.

Technology funding

Community Technology Foundation of California

http://www.zerodivide.org

Incorporated in 1998, the Community Technology Foundation of California (CTFC) promotes social justice, access, and equity through community technology. CTFC grants are made possible through the Community Technology Fund, and funding partnerships with the Marguerite Casey Foundation and the California Endowment. The Community Technology Fund is a ten-year, $50-million fund of CTFC established from the merger of SBC and Pacific Bell. Its resources are directed toward underserved communities in California.

"Developing a Technology Plan: Key to Getting Needed Funds"

http://www.uwnyc.org/technews/v3_n5_a1.html

This is a direct link to an article written by Radha Pillai, manager, Agency Relations, United Way of New York City.

The Foundation Center

http://www.foundationcenter.org

The Foundation Center's mission is to strengthen the nonprofit sector by advancing knowledge about U.S. philanthropy. The center offers funding and nonprofit resource libraries in New York, Atlanta, San Francisco, Cleveland, and Washington, DC, that you can visit. The web site offers information and links to grants and grantmakers.

Social Return on Investment (SROI) Collection

http://www.redf.org/results-sroi.htm

This set of publications captures REDF's efforts to calculate the SROI within its portfolio of social purpose enterprises. The SROI framework is one attempt to analyze and describe the impact of the enterprises on the lives of individuals and on the communities in which they live.

Tips, newsletters, and e-mail discussion groups

Apple Support Update

http://www.info.apple.com/subscribe/index.html

A newsletter from Apple Computer that includes security updates and other software changes and announcements.

Dummies eTips

http://etips.dummies.com/

The *Dummies* books are some of our favorites—and the tips are good too!

Global eRiders Discussion List

http://npogroups.org/lists/info/interider

A list set up to aid discussion among global eRiders and other nonprofit technology assistance providers. The primary language of the list is English. Traffic on the list is usually five to ten messages per week. Appropriate postings to the list include news, technical questions, and requests for assistance.

Langalist

http://www.langa.com

A newsletter that provides general computing information on topics such as Windows, Linux, firewalls, notice of important patches, and news about patches that might not work as advertised.

Microsoft Office Tips

http://office.microsoft.com/en-us/assistance/default.aspx

It's not a newsletter, but it's a good place for Office 2000 and XP tips.

Nonprofit Techie E-Mail List

http://lists.compasspoint.org/lists/info/npo-techies

One of the best tech resources around (if we say so ourselves)! This is an e-mail discussion list where you can post general or specific technical questions and get answers from your fellow nonprofit techies.

PC Show and Tell

http://www.pcshowandtell.com

An electronic library of thousands of tutorials that "show" and "tell" you, step-by-step, how to use and maximize popular products for today's electronic world. The site has over 35,000 "shows" on over 100 applications. Point your users in this direction if they need to know how to do something.

S'more Access

http://lists.compasspoint.org/lists/info/smore-access

Compasspoint runs the free e-mail list S'more Access, which is intended for members of the nonprofit community who are using and/or designing Access databases. Members are invited to post questions and share solutions to technical issues dealing with Access database development.

TechRepublic

http://www.techrepublic.com

You have to sign up for TechRepublic, but it's free and offers really good tips.

Information technology policies

ENTECH

http://epic.cuir.uwm.edu/entech/index.php

ENTECH began as the technology initiative of the Nonprofit Management Fund (http://www.nonprofitmanagementfund.org/enter.html), a foundation located in Milwaukee. The technology initiative was created to perform proactive, independent assessments of nonprofits' technology needs, particularly in the context of a grant request to improve the technical capacity of the organization. ENTECHcreated an IT policy template that generates a customizable document.

ASP (application service provider)

Antharia

http://www.antharia.com/

Content management (cms), contact management (crm) , marketing for nonprofits.

Convio

http://www.convio.com/

This site offers nonprofits ways to build strong constituent relationships online.

"The eNonprofit: A Guide to ASPs, Internet Services, and Online Software"

http://www.compasspoint.org/enonprofit/

This guide, written by Michael Stein and John Kenyon, offers a comprehensive look at application service providers for nonprofit users. It includes a beginner's guide to ASPs, a directory of links to over 100 providers, a detailed guide to selecting an ASP, tips on how to plan for successful implementation, and a collection of nonprofit Internet technology resources.

GetActive

http://www.getactive.com/

This site offers online communication tools for membership organizations. The GetActive suite consists of a central member management database and e-mail messaging module, along with complementary modules for grassroots advocacy, online fundraising, online community building, events management, and web site development and management.

Kintera

http://www.kintera.com/

Donor management and event fundraising ASP for nonprofits.

Nonprofit Matrix

http://www.nonprofitmatrix.com/

An online guide to commercial ASPs and portal providers for the nonprofit sector.

TechSoup

http://techsoup.org/techfinder/techsoup/index.cfm?P=simple_search_results&keywords=application+service+provider&zip=&range=%3c%3d10&submit=search

Resource list with ASP evaluations.

Volunteer Match

http://www.volunteermatch.org

Volunteer services-related ASP and portal provider.

Donor management software

Community TechKnowledge Online Solution
http://www.communitytech.net/
877-441-2111, Austin, TX

Allows nonprofits to manage, share, and analyze client, program, or funding data for accurate outcomes reporting. The software is web-based, including all of the system administration and reporting tools.

Donor Perfect by Starkland Systems
http://www.donorperfect.com
800-220-8111, Fort Washington, PA

Fundraising software for donor relations management in both PC/network and online web-browser-based systems.

DonorWorks by Starsoft Technologies, Inc.
http://www.starsoft.com
800-327-1476, Spokane, WA

Comprehensive constituent relationship management software that tracks your constituents and their relationship to your organization.

eBase by Groundspring.org
http://www.ebase.org
415-561-7807, San Francisco, CA

A complete nonprofit database solution for building and maintaining key relationships with organizational stakeholders (requires FileMaker Pro and Microsoft Office). Free to download but requires an annual membership fee.

"Evaluating Donor Management Software"
http://www.uwnyc.org/technews/pf_v7_n3_a2.html

A link to an overview article written by Arthur Vincie on nonprofit donor management needs, and a discussion about some current providers.

Exceed! by Telosa
http://www.telosa.com
800-676-5831, Palo Alto CA

Telosa's Exceed! Premier and Exceed! Basic fundraising and donor management software solutions meet the broad range of nonprofit needs and budgets.

Giftmaker Pro by Campagne Associates
http://www.campagne.com
800-582-3489, Manchester, NH

A donor tracking software that is offered in various versions to meet an organization's unique needs.

resultsplus by Metafile Information Systems
http://rp.metafile.com
800-638-2445, Rochester, MN

Windows-based fundraising software to manage constituents and time, as well as donations, mailings, reports, and special events.

Sage Fundraising 50 (formerly Paradigm)
http://www.mip.com/products/fundraising/paradigm/default.aspx
800-647-3863, Austin, TX

A complete set of core fundraising development program tools.

Accessibility and ergonomics

Accessible Internet

http://www.accessibleinter.net/

Focuses on assessing the accessibility of web sites and offers consultation on building or retrofitting sites to be as accessibility friendly as possible.

Alliance for Technology Access

http://www.ataccess.org

ATA and its members provide consulting on assistive technology and making web sites, technology, and programs accessible for people with disabilities and functional limitations.

A-Prompt

http://aprompt.snow.utoronto.ca/

Verifies that web pages are accessible to all people.

Disability & HR: Tips for HR Professionals

http://www.ilr.cornell.edu/ped/hr_tips/home.cfm

Articles, checklists, a glossary, and links to useful disability resources to help HR professionals be in accordance with the Americans with Disabilities Act (ADA).

JAN's Web Site Portal for Employers

http://www.jan.wvu.edu/

The Job Accommodation Network is a free service of the Office of Disability Employment Policy of the U.S. Department of Labor. It provides information on job accommodations, self-employment, and small business opportunities and related subjects.

Linux Accessibility Resource Site

http://lars.atrc.utoronto.ca/

As Linux gains ground in the desktop computing environment, accessibility becomes a key issue, just as in traditional software environments. Text-to-speech, high-contrast graphics, braille support, optical character recognition, and keyboard-mouse-input adaptations are all active projects that allow all users to compute effectively with today's free software. Visit this site for more information about the annual Linux accessibility conference and associated projects.

Microsoft's Accessibility Pages

http://www.microsoft.com/enable/

Guide to accessibility features of Microsoft products.

Section 508

http://www.section508.gov/

A web site that federal employees and the public can use to access resources for understanding and implementing Section 508—the law that requires that federal agencies' electronic and information technology be accessible to people with disabilities.

U.S. Department of Labor Occupational Safety & Health Administration eTool on Computer Workstations

http://www.osha.gov/sltc/etools/computer-workstations/

This e-tool illustrates simple, inexpensive principles to help create a safe and comfortable computer workstation.

Web Accessibility Initiative (WAI)

http://www.w3.org/wai/

WAI, in coordination with organizations around the world, pursues accessibility of the web through five primary areas of work: technology, guidelines, tools, education and outreach, and research and development.

WebXACT

http://webxact.watchfire.com

A free web service that lets you test single pages of web content for quality, accessibility, and privacy issues.

Security

Adware and spyware scanners

Ad-Aware

http://www.lavasoft.de

Software to find and remove adware and spyware from your computer.

Counter Spy

http://www.sunbeltsoftware.com

Software to find and remove adware & spyware from your computer

Pest Patrol

http://www.pestpatrol.com

Software to find and remove adware and spyware from your computer. This site also offers a free online scan that doesn't do removal.

Spybot Search & Destroy

http://www.safer-networking.org

Software to find and remove adware and spyware from your computer.

Spy Sweeper

http://www.webroot.com/

Software to find and remove adware & spyware from your computer.

Viruses and backups

AVG Antivirus

http://www.grisoft.com

Virus scanning software for Windows and Linux. Different versions are available for file and mail servers, and in a network version.

F-Prot Antivirus

http://www.f-prot.com

Virus scanning software for Windows, DOS, Solaris, Linux, and others.

F-Secure Antivirus

http://www.f-secure.com

Virus scanning software for Windows.

McAfee

http://www.mcafee.com

Virus scanning software for Windows computers. The site includes additional information on viruses and offers a free virus scan (but not removal). McAfee offers other security software as well.

Symantec

http://www.symantec.com/avcenter/

The security response page of Norton Anti-Virus. Virus scanning software for Windows and Mac OS. The site includes additional information on viruses & virus hoaxes. A selection of other security software is also available, as well as a complete suite adding antispyware and firewall protection

Firewalls

Firewall Reviews

http://www.consumersearch.com/www/software/firewalls/fullstory.html

A summary of multiple reviews of firewalls.

How a firewall works

http://www.howstuffworks.com/firewall.htm

How a firewall works, and why a firewall is important to you.

Minitutorials.com

http://minitutorials.com/firewalls/fireindex.shtml

A short tutorial on firewalls.

Tinysoftware

http://www.tinysoftware.com

A link to TinyFirewall software for Windows computers.

Zone Labs

http://www.zonealarm.com

Zone Alarm firewall software for Windows computers.

Articles and tutorials on security

Intranet Journal

http://www.intranetjournal.com

Newsletters, articles, and other information on networks, compliance, security, and much more.

The CERT® Coordination Center FAQ

http://www.cert.org/security-improvement/#practices

Recommended steps to secure computer systems.

Security bulletins

Microsoft Security Bulletins

http://www.microsoft.com/security/bulletins/default.mspx

Sign up here to be notified of the latest Windows security vulnerabilities and patches.

USCERT—Computer Emergency Readiness Team

http://www.us-cert.gov

Reports of new viruses, software vulnerabilities, and security-related activity. Sign up on this web site for free e-mail alerts.

Open source software

Alternatives to Microsoft

http://www.kmfms.com/alternatives.html

A personal web page that recommends software alternatives (much of which is open source) to Microsoft products.

Just Say No to Microsoft

http://microsoft.toddverbeek.com

A techie's personal web page reviewing alternative software applications.

Open Source Initiative

http://www.opensource.org

OSI is a nonprofit corporation dedicated to managing and promoting the open source definition for the good of the community, specifically through the OSI-certified open source software certification mark and program. Read about successful software products that have these properties, and about the certification mark and program that ensures that software really is "open source."

Networking resources

Networking

Broadband Reports

http://www.broadbandreports.com

Provides information and communication on the subject of residential and small business broadband connections—DSL, cable, and other high-speed Internet services.

Macwindows

http://www.macwindows.com

A wonderful resource for accidental techies who maintain a network with both Macs and PCs.

Practically Networked

http://www.practicallynetworked.com

A great site about sharing a single network connection on your network. It has reviews of equipment and some how-to information.

Search Networking

http://searchnetworking.techtarget.com

A good general information site about networking.

World of Windows Networking

http://www.windowsnetworking.com

Learn about the mysteries (and there are lots of them) of Windows networking.

Network protocol analyzers

Commview

http://www.tamos.com/products/commview/

A network packet sniffer[11] (for Windows) for advanced troubleshooting of network problems. It monitors Internet and local area network (LAN) activity, can capture and analyze network packets, and has some good reporting functions.

Ethereal

http://www.ethereal.com

A free network packet sniffer for advanced troubleshooting of network problems. Available for numerous platforms, including Windows, Linux, Solaris, Mac OS, BEOS. Works with many connection types, including Ethernet, token ring, PPP, and others. See the web site for more information.

---

[11] A "sniffer" is program or device that monitors data traveling over a network.

Software & hardware resources

Software help

Idealware

http://www.idealware.org

Idealware provides candid reviews and articles about software of interest to nonprofits, centralized into a web site. Through product comparisons, recommendations, case studies, and software news, Idealware allows nonprofits to make the software decisions that will help them be more effective.

Microsoft Technet

http://www.microsoft.com/technet

This is the Microsoft site designed for techies. It has lots of information about their software products, bugs, updates, and more. A searchable knowledge base helps you to figure out those odd errors that you keep getting. Tutorials and white papers that may be of some help are also available.

Woody's Office Portal

http://www.wopr.com

A good site for information about Microsoft office tips, tricks, and problems. Sign up for an e-mail newsletter about Office and Windows that is very helpful.

Hardware help

Belarc

http://www.belarc.com

The Belarc Advisor builds a detailed profile of installed software and hardware, including Microsoft hotfixes, and displays the results in a web browser. All PC profile information is kept private on the PC and is not sent to any web server.

Macfixit

http://www.macfixit.com

A good starting point when your Mac is acting up.

PCMechanic

http://www.pcmech.com

This site offers lots of information about PC hardware. You can even learn to build your own PC at this site.

Tidbits

http://www.tidbits.com

Another good site for Mac information that runs a very popular newsletter for Mac users and also discussion boards. All newsletters are archived and indexed on this site.

UPS Selector

http://www.apcc.com/template/size/apc/index.cfm

Helps you select the size and type of UPS (uninterruptible power supply) for your system.

User support

Netopia eCare

http://www.netopia.com/software/products/ecare/

Netopia eCare 4.0 connects support agents and their customers with a live collaboration link. There is a very small install package.

PCAnywhere

http://www.symantec.com/small_business/
products/remote_pc_fax/pca12/index.html

Software that offers remote control capability, combined with remote management and file transfer capabilities, which helps to quickly resolve help desk and server support issues. Requires product install on both machines.

Remote Desktop

http://www.microsoft.com/windowsxp/
using/mobility/getstarted/remoteintro.mspx

Windows built-in tool that allows you to remotely control your computer from another office, from home, or while traveling. You can use the data, applications, and network resources that are on your office computer without being in your office.

Virtual Network Computing (VNC)

http://www.realvnc.com

VNC software makes it possible to view and fully interact with one computer from any other computer or mobile device anywhere on the Internet. VNC software is cross-platform, allowing remote control between different types of computers. For ultimate simplicity, a Java viewer allows remote control of any desktop from within a browser without having to install software.

Web site resources

Web site registration and hosting

Grassroots.org

http://grassroots.org/do/Home

Free web hosting for qualified nonprofits. Qualified organizations include nonreligious organizations involved in education, environmentalism, humanitarian relief, fighting disease, homelessness, crime control, political freedom, government reform, consumer protection, youth issues, and other like-minded causes.

Internet Corporation for Assigned Names and Numbers (ICANN)

http://www.icann.org/registrars/
accredited-list.html

A complete list of accredited domain registrars.

"Search Engine Optimization"

http://hotwired.lycos.com/
webmonkey/99/31/index1a.html

An article on how to send more search engine traffic to your site.

"Web Hosting for under Ten Bucks"

http://hotwired.lycos.com/
webmonkey/02/01/index4a.html

An article on affordable web hosting options.

Web site compatibility

Anybrowser.com

http://www.anybrowser.com/screensizetest.html

Allows you to check you web site's compatibility with a variety of web browsers.

Web standards

DMX Zone

http://www.dmxzone.com

Resources and web design standards for Dreamweaver MX users/designers.

Web Standards Project

http://www.webstandards.org/

A grassroots coalition fighting for standards that ensure simple, affordable access to web technologies for all.

Web traffic analysis and statistics

Analog

http://www.analog.cx

A service for analyzing web traffic. Offers free software.

Google's Zeitgeist Page

http://www.google.com/press/zeitgeist.html

Google web search trends and patterns updated on a weekly basis.

Web Trends

http://www.webtrends.com

A service that allows users to analyze their web site traffic and trends.

Developer web sites

Content Management Systems

http://en.wikipedia.org/wiki/List_of_content_management_systems

This is a list of content management systems that are used to organize and facilitate collaborative content creation. Many of them are built on top of separate content management frameworks.

Macromedia

http://www.macromedia.com

All things Macromedia-related, including free trials of cool tools.

Zeldman.com

http://www.zeldman.com

A newsletter/resource site on web design news and information and web standards.

Webmonkey

http://www.webmonkey.com

All things interesting and useful for webbies.

HTML tutorials and resources

The "Extra Resources" Site for HTML for the World Wide Web, Fifth Edition with XHTML and CSS: Visual Quickstart Guide

http://www.cookwood.com/html/extras/

A good book on HTML and its online companion resource page.

HTML E-Mail Suggestions

http://www.macromedia.com/cfusion/
knowledgebase/index.cfm?Id=tn_15200

Tips from Macromedia on how to use Dream-weaver to send HTML e-mail.

HTML Validator

http://validator.w3.org

A free service that checks documents containing HTML and XHTML for conformance to W3C recommendations and other standards.

Webmonkey Tutorial

http://hotwired.lycos.com/webmonkey/
authoring/html_basics/index.html

Links to a series of articles on the basics of HTML web authoring.

W3Schools HTML Tutorials

http://www.w3schools.com/html/default.asp

A plethora of HTML tutorials.

General web site resources

HTML Writers Guild

http://www.hwg.org

This site offers a low-cost e-learning system, the Web Design Training Program, which covers a breadth of web work from HTML and Flash to Dreamweaver and Photoshop. The site also has a great resource page including lots of links.

Kristinlong.com

http://www.kristinlong.com/tools.html

A comprehensive listing of web tools for nonprofits from CompassPoint's "Accidental Webbie" faculty.

Good books for webbies

Building Accessible Websites by Joe Clark

New Riders Press, ©2002

http://www.joeclark.org/book/

Accessibility is the next frontier in web development. Making your web content available to everyone—including those with disabilities—is a growing concern. This book tells you everything you need to know about how to do this and then some.

Designing with Web Standards by Jeffrey Zeldman

New Riders Press, © 2003

http://zeldman.com/dwws/

A favorite web book. Funny and engaging, this book has done more to advance proper coding techniques and web standards practices than anything else printed on a page.

Eric Meyer on CSS by Eric Meyer

New Riders Press, © 2002

More Eric Meyer on CSS by Eric Meyer

New Riders Press, © 2004

http://www.ericmeyeroncss.com/

Great project-based tutorials on building web sites using CSS.

Search Engine Visibility by Shari Thurow

New Riders Press, © 2002

http://www.peachpit.com/title/0735712565

From planning and coding to linking campaigns and site popularity, this book will help you drive visitors to your site and put your site on the map.

The Unusually Useful Web Book by June Cohen

New Riders Press, © 2003

http://peachpit.com/title/0735712069

This book is a great overview of web site development from planning to maintenance.

Visual Quickstart Guides, Published by Peachpit Press

(various authors and publication dates)

http://www.peachpit.com/series/series.asp?Ser=335245

Peachpit makes Visual Quickstart Guides for all sorts of software as well as HTML and web programming languages. They are indispensable.

Usability[12]

Useit.com: Jakob Nielsen's web site

http://www.useit.com

Jakob Nielsen is the web's leading usabilty expert. Check out his web site for helpful usability ideas.

Nielsen Alertbox

http://www.useit.com/alertbox/

Sign up to join a great mailing list filled with ideas about making your web site more usable.

E-mail and integrated service providers

Groundspring.org

http://www.groundspring.org/index_gs.cfm

Groundspring.org is a nonprofit organization that provides simple, affordable, and integrated services for small-to-midsized nonprofit organizations to help them become effective users of Internet technology in their fundraising and management of donors and supporters.

NPO-mail and NPOGroups

http://electricembers.net/guide.php

Two components of an e-mail and listserver ASP that provides small-to-midsized nonprofits with services comparable to those of corporate in-house e-mail systems, without the corporate price tag or the hassles and disadvantages of maintaining a dedicated in-house server. A service of Electric Embers.

Mailing lists

Sparklist

http://www.sparklist.com

E-mail mailing list provider.

Topica

http://www.topica.com

E-mail mailing list provider.

Yahoo Groups

http://groups.yahoo.com

E-mail mailing list provider.

---

[12] A usable system (software, web, or other) is one that enables users to perform their job effectively and efficiently.

Fax services

eFax

http://www.efax.com/en/efax/twa/page/homepageplus

A service that allows you to fax via e-mail.

Fax Away

http://www.faxaway.com

A service that allows you to fax via e-mail.

Online surveys

Survey Monkey

http://www.surveymonkey.com/home.asp?Bhcd2=1060957381

Allows you to create professional online surveys quickly and easily.

Zoomerang

http://www.zoomerang.com/login/index.zgi

An Internet-based survey tool that allows customers to design and send surveys and analyze the results in real time.

Miscellaneous cool stuff

Google Toolbar

http://www.google.com/options/defaults.html

Directions on how you can make Google your default search engine, as well as other browser tricks.

Sitemap Generator

http://urlgreyhot.com/graphviz/

A web application that accepts uploaded tab-delimited text files and converts them into clickable site maps using Graphviz. The application was created to demonstrate how Graphviz can be used as a tool for information architecture work.

What Do I Know?

http://www.whatdoiknow.org

Articles on web design and development.

Yahoo Toolbar

http://companion.yahoo.com

Info on how to install a Yahoo toolbar that you can customize and access from any PC and search from anywhere on the web.

APPENDIX B

Taking Inventory

The following inventories can be downloaded at the publisher's web site at http://www.FieldstoneAlliance.org/worksheets. Simply use the code W490Ats05 (which is case sensitive) to download the materials. If you have any difficulties, phone the publisher at 800-274-6024. The materials are also available for download at www.AccidentalTechies.org.

Workstation Inventory

Purchase information (if available)

New:	If donation, from whom:	Date purchased:	Accepted by:
Vendor/Invoice:			
Phone:	E-mail:		Contact:
Warranty:	Term:		Warranty expires:
Tech support phone:			
Comments:			

(continued)

Inventory information

Inventory Date:	Make/model	ID no./serial no.	Cost
__Tower __Laptop __Desktop			
Monitor (e.g., Dell M90 15" CRT, 17" ViewSonic VE175 LCD)			
Processor type/speed (e.g., Pentium4, 1G, Celeron, AMD)			
Hard disk size (e.g., 20 GB)			
Removable storage (e.g., CD-RW, CD, floppy and/or Zip)			
Docking station			
Modem __internal __external			
Scanner			
Local printer			
Operating system (e.g., Windows 9x, 2000, OSX)			
Latest OS patch (date or version)			
USB ports (number, location)			
Assistive technology			
Other:			

Network information

Network card (Model, driver, connection 10/100/G)	
Office location/network ID	
Network printers attached	

(continued)

Software on this workstation

Type	Name	Version	CD Key
Word processor			
Spreadsheet			
Desktop publishing			
Database			
E-mail			
Web browser			
Backup			
Antivirus			
Update method			
Assistive technology			
Other:			

Staff who use this workstation

Date	Name	Dept.	Phone	E-mail	Office

Service and problem reports

Date	Resolved	By	Details

Replacement and retirement information

Date	Part Replaced	Comments	Destination

User Inventory

Employee ID	Name	Title	DOH	Department	Supervisor	Primary Workstation

Operating System	Location	Non-Std Apps	Std Apps	Non-Std Equip	Trainings Attended	Training needed

Database Planning

Step-by-step Planning

Step 1: Create a report list

The best way to start the planning process is to make a list of all the reports that you would like to generate from the database. To do this

- Gather any paper reports that your organization generates.

- Meet with staff to determine reporting needs that aren't currently being met.

- Create a *report list*. Give each report a descriptive name and purpose, and then list needed fields,[13] criteria, and statistics. A sample is below.

Report 1	
Name of report	*Big donors*
Purpose of report	*List used by executive director and board to solicit donations*
Fields needed	*First name, last name, phone numbers, e-mail, pledge amounts, and dates*
Criteria for selecting records for this report	*Select by date of donation, donations over $500, paid = yes*
Statistics needed	*Sum of donations by donor*

Report 2	
Name of report	*Client ethnicity by program*
Purpose of report	*Sent to funders each quarter*
Fields needed	*Client ID, ethnicity, program, service date*
Criteria for selecting records for this report	*Select by service date ranges*
Statistics needed	*Count of clients for each ethnicity, percentage of each ethnicity*

---

[13] *Fields* are the categories in a database, such as first name, last name, donation amount, and so forth.

Step 2: Create a form list

Next you need to list the various forms that people use or need. To do this

- Gather any paper forms that your organization uses to collect data.

- Meet with staff to determine form needs that aren't currently being met.

- Create a form list. Give each form a descriptive name, and then list its purpose and fields. A sample is below.

Form 1	
Name of form	Pledge form
Purpose of form	Filled out and sent in with pledges
Fields needed	First name, last name, address, city, state, zip, pledge level, pledge amount, payment type, credit card number, credit card expiration date

Form 2	
Name of form	Client service slip
Purpose of form	Keep track of each client's services to enter into the database
Fields needed	Client ID, client first name, client last name, case manager first name, case manager last name, service date, service type, service hours

Step 3: Create a comprehensive field list

Once you have finished listing report and form needs, use them to create a comprehensive list of all the fields that you will need in the database. At this point you are just determining database needs, not actually designing the database from a technical standpoint, so you do not need to worry about structuring the actual data tables. This part of the process is just about determining what fields are needed.

Drop-down lists

Some fields in the database will use drop-down lists to speed up data entry and to make the stored data consistent. For example, a case manager field would probably use a drop-down list of all the case managers at the organization, and an ethnicity field would use a list of ethnicities. When creating your comprehensive field list, it is good to specify the drop-down lists too.

Formatting

If you have specific formatting needs for some fields, list those here.

Using the reports and forms listed in the last two worksheets, start creating a comprehensive field list for the database, including formatting needs and drop-down list needs when appropriate. A sample follows.

Field name	Formatting needs or drop-down list (if appropriate)
Client first name	
Client last name	
Client home phone	555-555–5555, ext. 5555
Service date	
Service type	Drop-down list: individual counseling, group, home visit
Client ethnicity	Drop-down list: African American, Asian, Latino/a, Native American, Caucasian

Step 4: Diagram office procedures

A database is supposed to make data collection easier, not more difficult. To design a user-friendly database, you need to understand in detail the data collection and data entry procedures at your organization. The data entry forms (screens) need to be designed to allow for efficient data entry.

To understand the data flow in your office, it can be helpful to create diagrams showing how the data will get to the database. For example, here is a diagram from an imaginary organization offering counseling services to clients. This diagram maps the process of a new client entering the system. A sample is below.

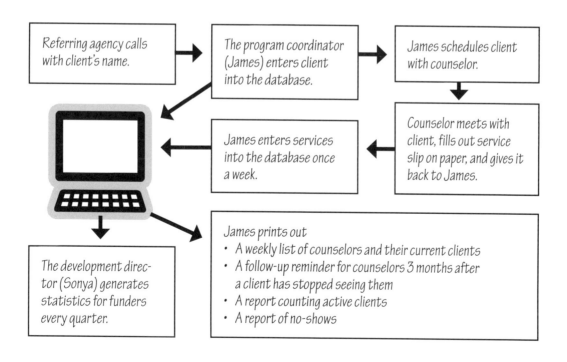

Referring agency calls with client's name.

The program coordinator (James) enters client into the database.

James schedules client with counselor.

Counselor meets with client, fills out service slip on paper, and gives it back to James.

James enters services into the database once a week.

The development director (Sonya) generates statistics for funders every quarter.

James prints out
- A weekly list of counselors and their current clients
- A follow-up reminder for counselors 3 months after a client has stopped seeing them
- A report counting active clients
- A report of no-shows

You might need to create a number of diagrams to capture various procedures that will involve the database. For example, at the same organization, when someone sends in a donation, Sarah, the administrative assistant, needs to check to see whether that person is already entered into the database as an organizational contact. If not, she enters the contact information and then enters the pledge. The data flow looks like this:

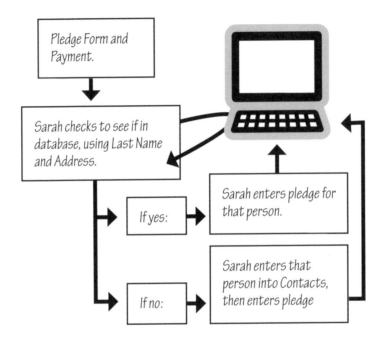

Sarah usually only gets a few pledges a week, but once a year, during the annual fund drive, she receives hundreds at a time, and needs to enter those in batches rather than having to first go to each contact's record.

It is also important to note that Sarah doesn't get along with Sonya, the development director. Sonya currently keeps her own records about donations in Excel. In cases like this, you will need to think about how personality problems between employees will hinder the effectiveness of a centralized database.

Office procedures and the database

Why is it important to understand the data flow in your office when planning a database? Because the database needs to fit properly into the flow rather than disrupting it. The data flow informs

• What data entry screens will be needed, and how they should look

- What navigation between data entry screens will be needed

- When and how reports will need to be printed out

- How many staff members need access to the database

- What security levels will be needed for staff members (in terms of their ability to change the data versus just being able to print out reports)

- What changes in procedures will be needed to get the data into the database

- What changes in procedures would make the data flow more efficient

- What interpersonal problems in the office need to be dealt with to have a centralized database

While examining office procedures, you will often find problems in the current data collection system. The database planning process is a time to clean up problematic office procedures.

The following pages are blank worksheets to help you plan or evaluate a database.

Database planning worksheets

Step 1: Create a report list

Report 1

Name of report	
Purpose of report	
Fields needed	
Criteria for selecting records for this report	
Statistics needed	

Report 2

Name of report	
Purpose of report	
Fields needed	
Criteria for selecting records for this report	
Statistics needed	

Report 3

Name of report	
Purpose of report	
Fields needed	
Criteria for selecting records for this report	
Statistics needed	

Database planning worksheets (continued)

Step 2: Create a form list

Form 1	
Name of form	
Purpose of form	
Fields needed	

Form 2	
Name of form	
Purpose of form	
Fields needed	

Form 3	
Name of form	
Purpose of form	
Fields needed	

Form 4	
Name of form	
Purpose of form	
Fields needed	

Database planning worksheets (continued)

Step 3: Create a comprehensive field list

Field name	Formatting needs or drop-down list (if appropriate)

Database planning worksheets (continued)

Step 4: Diagram office procedures

Security Policy Checklist [14]

Note: This list is meant primarily as a memory aid and should not be considered as complete instructions on how to write a security policy.

Management

1. What information technology security, use (including Internet and e-mail policies), and privacy policies are in place at the organization?

2. Are staff and management fully aware of these policies?

3. Are there any enforcement procedures?

4. Does someone in senior management have oversight for (or close involvement in) information technology security?

5. Are procedures or guidelines specified for acquiring and installing LAN peripherals, accessories, and so forth?

6. Is there a procedure to formally report security breaches?

7. Are emergency and disaster procedures established with well-defined tasks and responsibilities?

8. Is a proper backup plan in place so that operations can return to normal in a case where an installation causes problems? Is this plan tested?

9. Does the organization's computer software usage policy prohibit software piracy? Are users informed of the policy?

10. Are there proper inventory controls for the software and hardware?

11. Do users know who is in charge of information technology security, how to get in touch with that person, and when such contact is required?

12. Is it clear that the organization is in compliance with laws that apply to data security and privacy? (for example, the Health Insurance Portability and Privacy Act [HIPPA])

Connections from the Internet

1. What computers should be accessible to or seen by other computers from the outside? What services on those computers should be visible?

---

[14] This checklist was adapted from a checklist written by Brion Moss. © 1997 Brion Moss.

2. If computers are outside the firewall, how well are they separated from the inside? Could those computers be used to gain access to computers inside the firewall?

3. What information is allowed on computers outside the firewall?

4. Are connections from the Internet to the internal network allowed? How are they authenticated? Are they encrypted?

5. Who has access to externally visible computers? Who can make changes to them? (for example, Who can configure the firewall? Who can install scripts on the web server?)

Connections to the Internet

1. What traffic is allowed to go outside the internal network?

2. If there is traffic across the Internet between offices, how is it secured?

3. What protection is in place against viruses? against hostile Java applets?

Dial-up connections

1. Are dial-up connections allowed?

2. How are they authenticated?

3. Is wiretapping considered a threat? If so, how is it addressed?

4. What level of access to the internal network do dial-up connections provide? How does that compare to the access they should provide?

5. How are modems distributed? Can an employee set up a modem connection to their desktop computer?

Physical security

1. Are all systems physically protected from outsiders?

2. Is important equipment adequately secured from insiders?

Passwords

1. Are reusable passwords ever used externally? internally?

2. Are employees forced to change passwords periodically?

3. Are any efforts made to ensure that employees choose strong passwords?

4. If an unauthorized person gets a user's, what can they do with it? Are there additional barriers to entry? (for example, hardware keys or host filtering)

User rights and responsibilities

1. What systems should each user have access to (physically and virtually)? Is access to critical systems restricted? (for example, Are network servers in a locked room and only designated users can log into them?)

2. How much administrative authority are users allowed on their workstations or on the network?

3. How much freedom are users allowed in their system? Can they choose their own OS? software? Are there limits such as restrictions on games or certain software packages that are considered "nonsecure"?

4. What are users allowed to do with their accounts? Can they send and receive personal e-mail? Can they do personal work on company equipment?

5. What guidelines exist regarding resource consumption? (for example, How much disk or CPU usage are users allowed? Are there quotas on how much Internet bandwidth users may consume?)

6. What consequences does a user face for abuse (accidental or otherwise) of available computer services? (for example, printers; mail, file, and fax servers; or web servers)

7. Are users allowed access to each other's files? Should access to files be restricted? Is the lack of restrictions implicit permission for any user to access any file?

8. Can users share accounts with other users? with friends?

9. Can a user take company information home? (for example, copy files on a floppy and take them home to work on them) How is information secured outside the building?

10. What happens to a user's account when the user leaves the company?

11. Who keeps track of user accounts, making sure access is given and revoked properly?

12. Who is granted access to computer resources? just employees? What about contractors, partner companies, vendors, volunteers, or friends?

13. Do users understand the policy regarding e-mail privacy? regarding inflammatory mail or postings? regarding mail forgery?

Administrator rights and responsibilities

1. When may an administrator examine a user's account? their e-mail? What parts may administrators examine and what parts are off-limits? (for example, looking at an employee's Internet Explorer file of favorite bookmarked web sites may be acceptable, while looking at the employee's Internet Explorer history file is not)

2. Can the administrator monitor network traffic? within what boundaries?

3. Who may approve new accounts and access to services?

4. Who may give out supervisor access? Is there a record of who has it?

Sensitive information

1. How is sensitive information protected online?

2. How are backup tapes protected?

Emergency procedures

1. What procedures are in place for installing security-related patches?

2. What procedures are in place for handling a break-in? (for example, Is the network immediately shut down, or is ther an attempt to monitor the intruder?)

3. How are employees notified of a break-in?

4. At what point are law enforcement agencies involved?

5. Is there a well-known, easily accessible document containing emergency contact information?

6. What procedures exist for when an employee with high security access leaves?

7. Who makes the call in an emergency? Is this defined?

Documentation

1. Is there a map of the network topology? Is it clear where each piece of equipment fits on that map?

2. Is there an inventory of all hardware and software?

3. Is there a document describing the desired security configuration of each system?

4. Are all of these policies written down and understood by all involved?

Backups

1. What systems are backed up? How often?

2. How are backups secured?

3. How often are backups verified?

Logs

1. What information is logged?

2. How and where is the information logged?

3. Are the logs secure from tampering?

4. Are the logs regularly examined? by whom? How are logs filtered?

5. How are logs protected from prying eyes?

Networking Terms Accidental Techies Need to Know

firewall: commonly used in the presence of dedicated connections to the Internet ("always on" connections like DSL, as opposed to dial-up access) to prevent unauthorized access from the Internet to the local network. A firewall can be built into the hardware of a router or function through the software on a server or workstation.

hub: a central connection point in a network. Think of it as a box that connects multiple computers and other devices (such as a printer) to the network.

IP (Internet Protocol) address: a number needed by each workstation in a network in order to connect to the Internet. This number serves as a unique identifier and allows each workstation to operate on the network without conflict. These addresses may be permanently assigned to each device or dynamically assigned by a server or router through a protocol called DHCP (Dynamic Host Configuration Protocol). Confused? Think of it this way: if you manually enter an IP address into a workstation on your network, it's using a static IP address. If you simply plug a new workstation into the network and it automatically connects to both the server and the Internet, then it's using DHCP.

NAT (Network Address Translation): a piece of software or hardware that converts an IP address from a private address (on your local network) to a public address (out on the Internet). This enables your users to share a single public IP address. It also acts as a simple firewall to prevent access to your network from the outside. NAT is often used in homes and businesses to allow multiple PCs to access the Internet via T1, DSL, or cable modems.

network: a way of connecting workstations together so that they can share resources such as an Internet connection, files, printers, scanners, and so forth. Offices today that have more than one computer often have a network that links the computers together. To be networked, each workstation must have a network card (either Ethernet or wireless), a method of connecting (cable or wireless), and a connection point (a hub or switch). Almost all computers purchased today are sold with an Ethernet card built in.

router: a piece of hardware or software used to direct network traffic between segments of a network or to connect your network to an Internet service provider (ISP) for Internet communications.

server: a computer that provides services to other computers on the network (defined above). These services can include Internet access, file storage, security, and printing.

switch: kind of a superhub, a switch also connects devices in a network, but while a hub can only support communication between two connections (or ports) at one time, a switch allows multiple connections between different pairs of ports. What does this mean in the real world? a much faster network.

workstation: in most contexts, this simply refers to a computer (PC or Mac) used by a staff member. Computer manufacturers use this term to refer to high-powered computers used for demanding applications such as engineering or video production.

Characteristics of Nonprofit Organizations[15]

Introduction

The work of nonprofit organizations touches everyone who lives in the United States. Our air is cleaner, our culture is enlivened, our freedoms are protected and enhanced, the poor and sick among us live better lives all because of the work of nonprofit organizations.

While as a group "nonprofit organizations" covers a wide spectrum of size, scope and sophistication, the vast majority of the over one million nonprofit organizations in the United States are small to medium size, with fewer than fifty staff, budgets of less than $5 million, and have a mission focused on service. The following are nine "key" characteristics of these organizations:

1. Passion for mission

2. Atmosphere of "scarcity"

3. Bias toward informality, participation, and consensus

4. Dual bottom lines: financial and mission

5. Program outcomes are difficult to assess

6. Governing board has both oversight and supporting roles

7. Third-party funding

8. Mixed skill levels of staff (management and program)

9. Participation of volunteers

These characteristics have been grouped into three levels—corresponding to the three levels of an organization. The broadest category is *organization culture,* the normative behaviors and assumptions that tend to pervade life in nonprofit organizations. The second category is *organization structure,* the aspects of how nonprofit organizations are put together. And the third category is *individuals,* the people who work in the nonprofit sector.

These characteristics will influence—sometimes helping and other times being a potential barrier to—an organization's ability to achieve its mission. It is important for board and staff to be able to understand these unique characteristics, how they may affect behavior, and what board and staff can do to take advantage of strengths and overcome weaknesses.

[15] By Mike Allison and Jude Kaye. © 1998 CompassPoint Nonprofit Services.

Characteristics of Organization Culture

1. Passion for mission

The passion for mission characteristic of nonprofit organizations is one of their greatest sources of strength. The institutionalized impulse to "change the world" has brought about many of the most important advancements in American society. As a strength, the passion for mission taps incredible creativity, energy, and dedication for the work of an organization. However, zeal for the mission can lead staff, board, and volunteers involved with nonprofits to discount business realities, to turn strategic differences into interpersonal conflict, and to work with an urgency that borders on a crisis mentality or that leads to burnout.

2. Atmosphere of "scarcity"

This characteristic has both a factual component and a perceptual component. Most nonprofit leaders could do more work to accomplish their missions if they had more money, more access to decision making, more talented board members, and so forth. They are often, in fact, "underresourced" relative to their optimum scale. The fact that money, especially, takes a lot of energy to acquire can lead to a hyper cost-consciousness. In addition, the altruistic orientation of the organization also often leads to a sense that "most of our resources should go to the clients." One implication is that small and medium-sized nonprofit organizations often have underdeveloped infrastructures. Another implication is that nonprofit staff are often more willing to spend time (their own, volunteers', board members') rather than money to hire additional staff or consultants to get work done.

3. Bias toward informality, participation, and consensus

The lack of attention to hierarchy, a sense of friendliness, and a welcoming atmosphere are often described as attractive dimensions of a nonprofit culture. On the other hand, taken too far, informality may limit the appropriate exercise of authority. Informality may blur boundaries between the roles of staff (paid and volunteer), board, constituencies, and so forth. Overparticipation can inhibit the appropriate division of labor; the tendency toward consensus can bog down decision making. It is important that assumptions regarding who will participate in various types of decisions be made explicit rather than remain implicit or unclear.

Characteristics of Organization Structure

4. Dual bottom lines: mission and financial

One can debate to what extent this characteristic is *unique*. For-profit organizations have increasingly focused on the importance of mission relative to the primacy of return on investment. Governmental organizations have increasingly focused on the importance of mission relative to the primacy of political impact. Nonetheless, the tension between mission and financial results is a fundamental characteristic of nonprofit organizations. Internally, management of this tension is a factor in many strategic decisions, in making sense of "how well the organization is doing," and at all operational levels. Externally, some stakeholders of a nonprofit care about financial bottom lines (funders, competitors, and regulators) and some stakeholders care primarily about mission (clients and community). By recognizing the importance of both financial bottom lines and mission, staff, board, and constituencies can work together to create organizations that are both socially relevant or meaningful and financially viable.

5. Program outcomes are difficult to assess

In addition to the complexity involved with dual bottom lines, most nonprofit organizations have limited program evaluation capacity. This limitation is both partially caused by and exacerbated by the lack of standardized program outcomes in most fields. In child care, for example, standards for adult-child ratios exist, but little is standardized in terms of the quality of care delivered. Similarly, arts groups, advocacy organizations, mental health agencies, and community development corporations face substantial challenges in measuring their effectiveness. One implication is that assessing cost-effectiveness, and thus comparison of alternative actions, is difficult. Another implication is that different individuals may make different *assumptions* about the relationship between cost and effectiveness. Finally, some groups essentially ignore the issue by assuming their efforts are as effective as they can be. Recognizing the

difficulty of assessing outcomes can be an important part of the conversation when making hard choices about which services or programs to offer, to not offer, or to cut.

6. Governing board has both oversight and supporting roles

The governing board of a nonprofit has dual roles: it is responsible for ensuring that the public interest is served by the organization and—unlike private sector or government boards—it is expected to *help* the organization be successful. The first role is analogous to protecting the interest of stockholders or voters. The second role complicates the distinction between governance and management because in this role board members often do stafflike work. As helpers board members may raise funds, send mailings, paint buildings, or do the bookkeeping. This can lead to confusion about when and how it is appropriate for board members to be involved in various activities. Board members need to remember when they are wearing their "governance" hat and when they are wearing their "volunteer helper hat."

Furthermore, board members are often not expert either in nonprofit management or in the organization's field of service. As a result, board members either may be unprepared to make governance decisions or may inappropriately give up their fiduciary authority to staff. Critically important is recognizing the fine line between effective governance and micromanagement—it is inappropriate for the board to micromanage. Ultimately, the board/staff partnership is critical for ensuring the success of an organization.

7. Third-party funding

Many nonprofit organizations rely on third-party funding—grants from foundations, government agencies, and corporations. Third-party funding provides necessary support for services that cannot generate (sufficient) fees from client population. However, this funding usually comes with strings attached—restrictions, excessive reporting requirements, and/or directives regarding which services the funder thinks are in the best interest of the community (or support the funder's own needs or focus). This can result in a "tail wagging the dog syndrome" where the funder drives the nonprofit's programs rather than the nonprofit being proactive and intentional about what services it should offer.

Characteristics of Individuals

8. Individuals have mixed skill level

As a function of passion for the mission, limited financial resources, and a shallow pool of candidates, nonprofit organizations often hire managers with limited management training and program staff with little program experience. And because these organizations are small, there is seldom much internal capacity to provide training for staff. Nonprofit organizations must commit resources to developing one of their most important resources—their staff.

9. Participation of volunteers

Many nonprofit organizations rely on the active participation of volunteers. These volunteers are the heart and soul of the organization. Members of the board of directors are normally not paid for their work. Other individuals contribute considerable time and effort in delivering services and providing administrative support. The contribution that volunteers make to the nonprofit sector is significant—indeed without volunteerism many needed social services would not be available to the public. However, volunteers usually come with their own issues: they have to juggle multiple commitments and the relative priority they assign to their volunteer job may have to be balanced with their paid job, family responsibilities, and other volunteer commitments. And certain volunteers may resent that some other people are being paid for work that they (the volunteers) are doing for free, feeling that everyone should be volunteering.

Mike Allison, MBA, is the former director of CompassPoint Nonprofit Services' Consulting and Research Group and a former nonprofit executive director. He consults on strategic planning, governance, organization development, and program evaluation.

Jude Kaye is a senior fellow with CompassPoint Nonprofit Services and a nationally respected author, trainer, facilitator, and consultant. She specializes in strategic planning and organization development.

Mike and Jude are the authors of the nationally respected book Strategic Planning for Nonprofit Organizations—A Practical Guide and Workbook, *published by John Wiley and Sons (now in its second edition).*

Note: A previous version of this article was first published in the Fall 1998 issue of Vision/Action: The Journal of the Bay Area Organization Development Network. *The original article was called "Seven Characteristics of Nonprofit Organizations." Right after the publication of that article, the authors added two additional (and very obvious) characteristics: participation of volunteers and third-party funding.*

 **Index**

d indicates diagram
s indicates sample worksheet
w indicates worksheet

More results-oriented books from Fieldstone Alliance

Management & Leadership

The Accidental Techie
Supporting, Managing, and Maximizing Your
Nonprofit's Technology
by Sue Bennett

How to support and manage technology on a day-to-day basis including setting up a help desk, developing an effective technology budget and implementation plan, working with consultants, handling viruses, creating a backup system, purchasing hardware and software, using donated hardware, creating a useful database, and more.

176 pages, softcover   Item # 069490

Benchmarking for Nonprofits
How to Measure, Manage, and Improve Results
by Jason Saul

This book defines a formal, systematic, and reliable way to benchmark (the ongoing process of measuring your organization against leaders), from preparing your organization to measuring performance and implementing best practices.

128 pages, softcover   Item # 069431

Consulting with Nonprofits: A Practitioner's Guide
by Carol A. Lukas

A step-by-step, comprehensive guide for consultants. Addresses the art of consulting, how to run your business, and much more. Also includes tips and anecdotes from thirty skilled consultants.

240 pages, softcover   Item # 069172

The Fieldstone Nonprofit Guide to
Crafting Effective Mission and Vision Statements
by Emil Angelica

Guides you through two six-step processes that result in a mission statement, vision statement, or both. Shows how a clarified mission and vision lead to more effective leadership, decisions, fundraising, and management. Includes tips, sample statements, and worksheets.

88 pages, softcover   Item # 06927X

The Fieldstone Nonprofit Guide to
Developing Effective Teams
by Beth Gilbertsen and Vijit Ramchandani

Helps you understand, start, and maintain a team. Provides tools and techniques for writing a mission statement, setting goals, conducting effective meetings, creating ground rules to manage team dynamics, making decisions in teams, creating project plans, and developing team spirit.

80 pages, softcover   Item # 069202

The Five Life Stages of Nonprofit Organizations
Where You Are, Where You're Going, and What to Expect When You Get There
by Judith Sharken Simon with J. Terence Donovan

Shows you what's "normal" for each development stage which helps you plan for transitions, stay on track, and avoid unnecessary struggles. This guide also includes The Nonprofit Life Stage Assessment to plot and understand your organization's progress in seven arenas of organization development.

128 pages, softcover   Item # 069229

The Manager's Guide to Program Evaluation
Planning, Contracting, and Managing for Useful Results
by Paul W. Mattessich, PhD

Explains how to plan and manage an evaluation that will help identify your organization's successes, share information with key audiences, and improve services.

96 pages, softcover   Item # 069385

The Nonprofit Mergers Workbook
The Leader's Guide to Considering, Negotiating, and Executing a Merger
by David La Piana

A merger can be a daunting and complex process. Save time, money, and untold frustration with this highly practical guide that makes the process manageable and controllable. Includes case studies, decision trees, twenty-two worksheets, checklists, tips, and complete step-by-step guidance from seeking partners to writing the merger agreement, and more.

240 pages, softcover   Item # 069210

The Nonprofit Mergers Workbook Part II
Unifying the Organization after a Merger
by La Piana Associates

Once the merger agreement is signed, the question becomes: How do we make this merger work? *Part II* helps you create a comprehensive plan to achieve *integration*—bringing together people, programs, processes, and systems from two (or more) organizations into a single, unified whole.

248 pages, includes CD-ROM   Item # 069415

Nonprofit Stewardship
A Better Way to Lead Your Mission-Based Organization
by Peter C. Brinckerhoff

The stewardship model of leadership can help your organization improve its mission capability by forcing you to keep your organization's mission foremost. It helps you do more good for more people.

272 pages, softcover   Item # 069423

Resolving Conflict in Nonprofit Organizations
The Leader's Guide to Finding Constructive Solutions
by Marion Peters Angelica

Helps you identify conflict, decide whether to intervene, uncover and deal with the true issues, and design and conduct a conflict resolution process. Includes guidance on handling unique conflicts such as harassment and discrimination.

192 pages, softcover   Item # 069164

Strategic Planning Workbook for Nonprofit Organizations, Revised and Updated
by Bryan Barry

Chart a wise course for your nonprofit's future. This time-tested workbook gives you practical step-by-step guidance, real-life examples, and one nonprofit's complete strategic plan.

144 pages, softcover   Item # 069075

Finance

Bookkeeping Basics
What Every Nonprofit Bookkeeper Needs to Know
by Debra L. Ruegg and Lisa M. Venkatrathnam

This book will enable you to successfully meet the basic bookkeeping requirements of your nonprofit organization—even if you have little or no formal accounting training.

128 pages, softcover   Item # 069296

Coping with Cutbacks
The Nonprofit Guide to Success When Times Are Tight
by Emil Angelica and Vincent Hyman

Shows you practical ways to involve business, government, and other nonprofits to solve problems together. Also includes 185 cutback strategies you can put to use right away.

128 pages, softcover   Item # 069091

Financial Leadership for Nonprofit Executives
Guiding Your Organization to Long-term Success
by Jeanne Peters and Elizabeth Schaffer

Provides executives with a practical guide to protecting and growing the assets of their organizations and with accomplishing as much mission as possible with those resources.

144 pages, softcover   Item # 06944X

Venture Forth! The Essential Guide to Starting a Moneymaking Business in Your Nonprofit Organization
by Rolfe Larson

The most complete guide on nonprofit business development. Building on the experience of dozens of organizations, this handbook gives you a time-tested approach for finding, testing, and launching a successful nonprofit business venture.

272 pages, softcover   Item # 069245

Marketing

The Fieldstone Nonprofit Guide to
Conducting Successful Focus Groups
by Judith Sharken Simon

Shows how to collect valuable information without a lot of money or special expertise. Using this proven technique, you'll get essential opinions and feedback to help you check out your assumptions, do better strategic planning, improve services or products, and more.

80 pages, softcover   Item # 069199

Marketing Workbook for Nonprofit Organizations Volume I: Develop the Plan
by Gary J. Stern

Don't just wish for results—get them! Here's how to create a straightforward, usable marketing plan. Includes the six Ps of Marketing, how to use them effectively, a sample marketing plan, tips on using the Internet, and worksheets.

208 pages, softcover   Item # 069253

For current prices, a catalog, or to order call ☎ 800-274-6024

Marketing Workbook for Nonprofit Organizations Volume II: Mobilize People for Marketing Success
by Gary J. Stern

Put together a successful promotional campaign based on the most persuasive tool of all: personal contact. Learn how to mobilize your entire organization, its staff, volunteers, and supporters in a focused, one-to-one marketing campaign. Comes with *Pocket Guide for Marketing Representatives.* In it, your marketing representatives can record key campaign messages and find motivational reminders.

192 pages, softcover   Item # 069105

Board Tools

The Best of the Board Café
Hands-on Solutions for Nonprofit Boards
by Jan Masaoka, CompassPoint Nonprofit Services

Gathers the most requested articles from the e-newsletter *Board Café.* You'll find a lively menu of ideas, information, opinions, news, and resources to help board members give and get the most out of their board service.

232 pages, softcover   Item # 069407

The Nonprofit Board Member's Guide to Lobbying and Advocacy
by Marcia Avner
96 pages, softcover   Item # 069393

Keeping the Peace
by Marion Peters Angelica

Written especially for board members and chief executives, this book is a step-by-step guide to ensure that everyone is treated fairly and a feasible solution is reached.

48 pages, softcover   Item # 860127

Funder's Guides

Community Visions, Community Solutions
Grantmaking for Comprehensive Impact
by Joseph A. Connor and Stephanie Kadel-Taras

Helps foundations, community funds, government agencies, and other grantmakers uncover a community's highest aspiration for itself, and support and sustain strategic efforts to get to workable solutions.

128 pages, softcover   Item # 06930X

A Funder's Guide to Evaluation: Leveraging Evaluation to Improve Nonprofit Effectiveness
by Peter York

Funders and nonprofit leaders are shifting away from proving something to someone else, and toward *improving* what they do so they can achieve their mission and share how they succeeded with others. This book includes strategies and tools to help grantmakers support and use evaluation as a capacity-building tool.

160 pages, softcover   Item # 069482

A Funder's Guide to Organizational Assessment
Tools, Processes, and Their Use in Building Capacity
by GEO

In this book, funders, grantees, and consultants will understand how organizational assessment can be used to build the capacity of nonprofits, enhance grantmaking, impact organizational systems, and measure foundation effectiveness.

216 pages, CD-ROM included   Item # 069539

Strengthening Nonprofit Performance
A Funder's Guide to Capacity Building
by Paul Connolly and Carol Lukas

This practical guide synthesizes the most recent capacity-building practice and research into a collection of strategies, steps, and examples that you can use to get started on or improve funding to strengthen nonprofit organizations.

176 pages, softcover   Item # 069377

Collaboration

Collaboration Handbook
Creating, Sustaining, and Enjoying the Journey
by Michael Winer and Karen Ray

Shows you how to get a collaboration going, set goals, determine everyone's roles, create an action plan, and evaluate the results. Includes helpful tips on how to avoid pitfalls and worksheets to keep everyone on track.

192 pages, softcover   Item # 069032

Collaboration: What Makes It Work, 2nd Ed.
by Paul Mattessich, PhD, Marta Murray-Close, BA, and Barbara Monsey, MPH

An in-depth review of current collaboration research. Includes twenty key factors influencing successful collaborations are identified. Includes The Wilder Collaboration Factors Inventory, which groups can use to assess their collaboration.

104 pages, softcover   Item # 069326

For current prices or to order visit us online at 🖥 www.FieldstoneAlliance.org

The Nimble Collaboration
Fine-Tuning Your Collaboration for Lasting Success
by Karen Ray

Shows you ways to make your existing collaboration more responsive, flexible, and productive. Provides three key strategies to help your collaboration respond quickly to changing environments and participants.

136 pages, softcover   Item # 069288

The Nonprofit Board Member's Guide to Lobbying and Advocacy
by Marcia Avner

Written specifically for board members, this guide helps organizations increase their impact on policy decisions. It reveals how board members can be involved in planning for and implementing successful lobbying efforts.

96 pages, softcover   Item # 069393

Lobbying & Advocacy

The Lobbying and Advocacy Handbook for Nonprofit Organizations
Shaping Public Policy at the State and Local Level
by Marcia Avner

A planning guide and resource for nonprofit organizations that want to influence issues that matter to them. This book will help you decide whether to lobby and then put plans in place to make it work.

240 pages, softcover   Item # 069261

Violence Prevention & Intervention

The Little Book of Peace

Journey Beyond Abuse: A Step-by-Step Guide to Facilitating Women's Domestic Abuse Groups

Foundations for Violence-Free Living: A Step-by-Step Guide to Facilitating Men's Domestic Abuse Groups

What Works in Preventing Rural Violence

ORDERING INFORMATION

Order online, or by phone or fax

Online: www.FieldstoneAlliance.org
E-mail: books@fieldstonealliance.org

Call toll-free: 800-274-6024
Internationally: 651-556-4509

Fax: 651-556-4517

Mail: Fieldstone Alliance
Publishing Center
60 Plato BLVD E, STE 150
St. Paul, MN 55107

Our NO-RISK guarantee

If you aren't completely satisfied with any book for any reason, simply send it back within 30 days for a full refund.

Pricing and discounts

For current prices and discounts, please visit our web site at www.FieldstoneAlliance.org or call toll free at 800-274-6024.

Quality assurance

We strive to make sure that all the books we publish are helpful and easy to use. Our major workbooks are tested and critiqued by experts before being published. Their comments help shape the final book and—we trust—make it more useful to you.

Visit us online

You'll find information about Fieldstone Alliance and more details on our books, such as table of contents, pricing, discounts, endorsements, and more, at www.FieldstoneAlliance.org.

Do you have a book idea?

Fieldstone Alliance seeks manuscripts and proposals for books in the fields of nonprofit management and community development. To get a copy of our author guidelines, please call us at 800-274-6024. You can also download them from our web site at www.FieldstoneAlliance.org.